T0329732

Research Design and Analysis

Research Design and Analysis

A Primer for the Non-Statistician

Leslie D. Rosenstein
University of Texas Southwestern Medical Center

The right of Leslie D. Rosenstein to be identified as the author of this work has been asserted in accordance with law.

Registered Office
John Wiley & Sons, Inc., 111 River Street, Hoboken, NJ 07030, USA

Editorial Office
111 River Street, Hoboken, NJ 07030, USA

For details of our global editorial offices, customer services, and more information about Wiley products visit us at www.wiley.com.

Wiley also publishes its books in a variety of electronic formats and by print-on-demand. Some content that appears in standard print versions of this book may not be available in other formats.

Library of Congress Cataloging-in-Publication Data
Names: Rosenstein, Leslie D., author.
Title: Research design and analysis : a primer for the non-statistician /
  Leslie D. Rosenstein, UT Southwestern Medical Center.
Description: Hoboken, NJ : Wiley, 2019. | Includes bibliographical references
  and index. |
Identifiers: LCCN 2019003539 (print) | LCCN 2019017653 (ebook) | ISBN
  9781119563624 (Adobe PDF) | ISBN 9781119563617 (ePub) | ISBN 9781119563594
  (hardback)
Subjects: LCSH: Medicine–Research–Methodology. | BISAC: SOCIAL SCIENCE /
  Sociology / General.
Classification: LCC R850 (ebook) | LCC R850 .R67 2019 (print) | DDC
  610.72–dc23
LC record available at https://lccn.loc.gov/2019003539

Cover image: © oxygen/Getty Images
Cover design by Wiley

Set in 10/12pt WarnockPro by SPi Global, Chennai, India

Printed in the United States of America

V10009944_050119

I am grateful for my loving and supportive family who has helped me overcome some hefty obstacles. I would not have been able to write this book without the support of Jean, Marv, Dana, Shari, Kevin, Cory, and Caleigh.

I also want to acknowledge my students at UT Southwestern Medical Center for their patience and interest. Their reactions during class discussions, though hard to read at times, helped guide me in organizing and formulating the chapters of this primer. I especially want to thank Dr. Mallory Jacobs who inspired me to try to write something succinct and user-friendly for busy physicians who want to be good consumers of clinical research.

Contents

List of Figures

List of Tables

Introduction

In this book, I set out to provide a hopefully, pain-free overview of research methods, design, and analysis. The intended audiences include those in the sciences who wish to conduct their own research without investing several semesters completing coursework in statistics and related fields, as well as those in the sciences, clinical fields, education, and the media who wish to read published research in an informed manner. In the former case, this manuscript will provide a general basis for designing and conducting research, though with the assistance of a statistical consultant. In the latter case, I hope this primer will provide a basis for reading, understanding, and critically evaluating research reports.

For health care providers who wish to read studies and make treatment recommendations to their patients based on study outcomes, I hope this book will be a good reference tool. Research publications can sometimes be full of nuances and jargon that are only meaningful to the trained researcher. Without a clear understanding of research design, validity, and interpretation, the results reported in publications can be misunderstood and applied improperly. Sometimes, the research may be poorly conducted or poorly reported, and a basic knowledge of research design and interpretation can be particularly useful in judging when that is the case. At other times, the research is well done, but difficult to understand without a basic knowledge of research methods.

Professionals working in the media are well aware of their great responsibility in reporting research findings to the public. The media has a special role in providing information to the public while avoiding harm as outlined in the *Professional Journalists' Code of Ethics* (Society of Professional Journalists, 2014). That code also mandates that journalists are responsible for the accuracy of their reporting, including verifying the information before it is released.

Carrie Figdor (2017) points out the difficulty presented to journalists in their role of reporting and providing information that is accurate when the material is the product of scientific endeavors. Journalists cannot necessarily rely on authors of scientific reports to provide accurate and valid information, and this quandary has become exponentially worse with the evolution of mass communication tools. Non-peer-reviewed research reports are more readily available to the masses. Moreover, journalists cannot necessarily count on peer-reviewed journals to publish only sound research. Most do, but journalists must be careful, yet, to review and understand the research design as presented along with the results and conclusions.

Journalists must take care, for instance, to not translate a conclusion of an association between two events or variables into a claim of causality. Oftentimes, the correct language to that effect is included in a research publication, but it is incumbent on the journalist to read and understand such language. Otherwise, there is a real and great risk that the public will be misinformed and harmed as a result. In Chapter 9, I discuss this in more detail with respect to specific instances of marked harm being perpetrated unintentionally (e.g. the unsubstantiated fear of the measles vaccine, misinformation about the true risks of chronic traumatic encephalopathy, and misinterpretation of the Women's Health Initiative findings).

The chapters of this book are laid out into four major sections. In Section 1, I briefly review the purpose of research as well as ethics and rules guiding research involving human participants and animal subjects. In Section 2, I walk you through basic research designs and validity. In Section 3, I provide a cursory review of statistical techniques, just enough to make you conversant with your statistical consultant or to be able to comprehend the jargon you find in many research documents. I have also included a chapter on meta-analytic studies. The goal of that chapter is to help you in sifting through reports of meta-analyses, though I also provide some direction in case you ever consider conducting your own meta-analytic study. In the fourth section, I review the how-tos of disseminating research findings, including reporting and presenting research results. I discuss how to prepare a research paper for submission to a peer-reviewed journal. I also talk about the concept of poster presentations and how to submit research more quickly for presentation at a conference.

In Section 4, I also present my concluding remarks. There, I repeat what I emphasize throughout this primer; that is, research and research findings are only as good as the research design. Most importantly, it is crucial to avoid making statements of claims of causality between two conditions, or variables, when the research design does not permit drawing such conclusions with any degree of confidence. Accurate interpretation of research findings is of critical importance. This does not just apply to the authors of the original research but also to others who report about and share research findings and claims more broadly. In particular, I hope to underline the importance and responsibility carried by journalists and others who discuss research claims. Sadly, when research claims are reported and shared with the public without a critical eye or with misstatements about causality, harm may ensue.

Finally, I have prepared appendices with tools for those who are planning to conduct their own research. These contain information about data sets, databases, statistical software programs, and resources for those who want to learn more about inferential statistics. I have additionally included a glossary of many of the terms included in this primer; in the glossary, terms are alphabetized for quick lookup.

Section 1

The Purpose, Ethics, and Rules of Research

1

The Purpose and Ethics of Research

1.1 The Purpose and Risks of Research

Why do we do research? There are many reasons: to answer a question, to advance understanding of a topic, to evaluate interventions, to predict behavior, to understand differences between groups, and so forth. When we conduct research, we usually start with an inquiry based on theory. We then develop hypotheses. Hypotheses are testable questions or predictions, which are ideally based on theory or pre-existing knowledge about a topic.

Is there ever a time when research should not be conducted? Yes! Logistically, some research ideas may not have benefits that outweigh the costs of conducting the research. These cost considerations include fiscal costs, time, and effort. But there are also ethical considerations in determining *whether* research should be conducted and *how* it is conducted.

In terms of the *whether*, one might ask if there are any potential harms of the research. Consideration of the issue of potential harm typically refers to the harm that may be incurred by the participants or subjects in the process of conducting the research, but harm can also theoretically result from the findings or the knowledge gained by the research. For instance, what if you want to know whether a necessary, life-saving treatment causes long-term cognitive impairments? One might argue that there is no point in "proving" the adverse effects of the treatment if the treatment is required for survival and there are no alternative options. On the other hand, perhaps patients deserve to be fully informed of the potential side effects before deciding whether to pursue treatment versus opting for fate. Perhaps, too, an understanding of negative side effects could lead to the development of strategies and interventions to minimize or reverse them.

As another example, many researchers may be interested in knowing the negative consequences of a certain type and severity of injury. If one stops there in the research, that is, if the research concludes following documentation of the negative impacts of an event or injury, then the research has

Research Design and Analysis: A Primer for the Non-Statistician, First Edition. Leslie D. Rosenstein.
© 2019 John Wiley & Sons, Inc. Published 2019 by John Wiley & Sons, Inc.

possibly not improved the human state. However, if used as a starting point, the findings from a line of research investigating those consequences could provide very important information, particularly if it opens the door to further investigations into the mechanisms of those consequences as well as future interventions to address them. In other words, if the next step is taken toward understanding the mechanisms that trigger those sequelae, that understanding can then be used to theorize about and then test potential therapeutic strategies.

The second set of ethical considerations referred to earlier pertains to the *how* research is conducted. If not carefully planned, research, particularly at the data-collection stage, can cause harm to the participating humans or animal subjects. Even when experimentation is carefully planned, harm may be inevitable. Harm can occur physically and/or psychologically through injury, misinformation, or misunderstanding.

At times, scientists may face a dilemma in asking whether their research will have more harm than benefit. There may be no clearly right or wrong answer to that question, but it is one that scientists should consider before pursuing a line of research. Ultimately, a researcher wants to be aware of all possible outcomes and impacts that their research may have, not just in the process of conducting the research but also in the process of sharing their results.

Regarding the potential harm to human participants and animal subjects in the process of experimentation, society is replete with examples of individuals and animals being injured physically and/or psychologically by scientific inquiries. This history and the rules and guidelines that have stemmed from attempts to prevent future harm are reviewed next.

1.2 History of Harm to Humans

I am hesitant to introduce the darker side of research in this primer that is, otherwise, intended to be a light overview of research methods. However, understanding harm inflicted on innocent individuals in the name of research throughout recent history puts into perspective the institutional research requirements most of us face when undertaking a new investigation. At times, working through the process of an Institutional Review Board (IRB; which I will discuss in Section 1.6) can be tedious and even frustrating. However, IRB members and other regulators are not trying to make research life difficult. While it does at times seem like we have to overcome irrelevant obstacles, these entities are tasked with the noble and critical goal of preventing a repeating of history. The following pages, therefore, are intended to provide perspective by reminding us of what has come before.

Cruelty inflicted on humans in experimentation actually dates back at least several centuries. In recorded history, convicted criminals were used in studies of human anatomy as far back as the fourth to third century BCE (see Franco (2013), for a review and discussion). Specifically, ancient Greek scientists performed vivisections and dissections on convicted criminals.

Probably the most egregious and well-known modern examples of unethical, horrific research experiments occurred during the reign and terror of the Nazi regime. Weindling, von Villiez, Loewenau, and Farron (2015) provide a systematic accounting of the atrocities inflicted on tens of thousands of adult and child victims from at least 24 countries during that period. There were a particularly large number of victims from Poland, including both Jewish and Catholic citizens. Many, many cases occurred in the context of imprisonment such as in concentration camps, while other cases occurred under coercion and without consent within psychiatric facilities. Many victims died or were killed in the course of experimentation, with some murders occurring in order to study bodies and body parts. Many other victims lived, but were left with serious, often life-altering injuries.

As outlined by Weindling et al. (2015), the Nazi experimentation began as an undertaking in eugenics. There, the goal of the Nazis was to wipe the land of individuals from different groups, including those of the Jewish faith, individuals identifying as Roma or Sinti, individuals of "mixed race," individuals with mental illness, and others. Experimentation grew from X-ray sterilization to also include studies in which victims were infected with diseases in order to test new drugs. Forced infection included diseases such as tetanus, typhoid, and typhus. Twins sometimes served as comparison controls for their victim sibling in these drug studies, and they were also used in gruesome experiments in which perpetrators attempted to conjoin twins surgically. There were experiments assessing the impact of exposure to high altitude/low pressures and/or extreme freezing temperatures. Anthropometric "data" were collected, including in the form of a "Jewish skeleton collection." In all, with just what is now known through others' tenacious inspection of records and interviews, there were a great number of atrocities in the name of experimentation committed against innocent, nonconsenting human beings during the reign of the Nazi terror.

At the same time that the Nazis were inflicting atrocities on humans as research guinea pigs, the Japanese Imperial Army and the Japanese Imperial Guard were, likewise, inflicting great harm on prisoners of war in the name of scientific investigation. Those atrocities have only come to be widely known in more recent years, however. Two major components of the wartime experiments occurred at Unit 731 and other prisoner-of-war camps in Japanese-occupied China, and in an American prisoner-of-war camp in Japan.

Unit 731 was a prisoner-of-war camp in Manchuria where Chinese inmates, both soldiers and locals, were subjected to horrific experiments as described by Herbert Kikoy (2018) and others. As reported by Kikoy, similar experiments were conducted on Russian and other allied war prisoners at nearby camps.

The perpetrators horrendously conducted live vivisections and other biological investigations. These included studies of biological weapons for warfare in which bombs were set off to test the ability to spread airborne infection of gangrene and other bacteria. Other similarities with the Nazi experiments were studies evaluating the impact of exposure to extreme temperatures or decompression. The Japanese Imperial scientists led by a surgeon also tested the effects of X-ray bombardment, starvation, and sleep deprivation in humans. In addition to the live vivisections, they boiled humans alive, and subjected others to centrifuges resulting in unimaginable terror, pain, and death.

In addition to the atrocities performed on the prisoners at Unit 731 and the nearby war camps, the Japanese Imperial research extended to studies of the plague involving the Chinese populace. The perpetrators bred rats that could be infested with fleas infected with the plague. The goal, again, was to develop biological weapons. Those experiments resulted in several outbreaks of the plague according to Kikoy (2018).

Another component of the Imperialists' atrocious research program occurred in Japan at the medical school of Kyushu Imperial University. One experiment involved injecting an anesthetized prisoner with seawater to test whether it could serve as a substitute for sterile saline solution (McCurry, 2015). According to McCurry, based on review of testimony used during war tribunals, organs were removed from prisoners to assess the impact of surgery on organ systems. One prisoner suffered through the experience of having his skull drilled through to test a surgical intervention for the treatment of epilepsy. Corpses of deceased prisoners of war were also preserved in formaldehyde for medical students to use in their studies. These and other details are discussed in a Japanese book written by a physician who was a young medical student at Kyushu Imperial University during this period; his accounts are reviewed, in English, by McCurry (2015) and others (e.g. O'Flynn, 2015).

Sadly, harmful experimentation inflicted on unknowing or unwilling individuals has also occurred outside of the context of war. In the United States, one of the most infamous and shameful of these experiments, originally known as the "Tuskegee Study of Untreated Syphilis in the Negro Male," was carried out by the United States Public Health Service. That study concluded in 1972 and involved monitoring the natural course of syphilis even after penicillin became the accepted treatment of choice in the 1940s (Centers for Disease Control and Prevention, 2017c).

As chronicled by the National Center for Bioethics in Research and Health Care (Tuskegee University, 2018), 600 men, including 399 who were found to have syphilis and 201 who served as a control group, were recruited into the study beginning in 1932. In exchange for their participation, they were offered, "medical exams, rides to and from the clinics, meals on examination days, free treatment for minor ailments and guarantees that provisions would be made after their deaths in terms of burial stipends paid to their survivors." The purpose of the study was to learn about the natural course of the disease. When the participants were initially recruited, there was not a proven treatment for syphilis. However, penicillin became the standard treatment in 1947 or earlier. In spite of this, treatment with the antibiotic was withheld in order to continue the study and the endeavor to learn more about the disease progression. The study was not halted until its existence was made public in a news story in 1972. Through a series of reviews and investigations instigated by the Assistant Secretary for the United States Health and Scientific Affairs, it was determined that the participants had never been adequately informed through a procedure called informed consent. Related, they were never warned of the risks of the disease to themselves, their sexual partners, or their children conceived after infection. They were also reportedly never given the choice to withdraw from the study in order to receive treatment once it became available. Ultimately, the study participants and their families were awarded a sum of more than $9 million as part of a class-action law suit, but the harm that had been inflicted could not be undone.

There are many examples throughout history of marginalized individuals being used in research without informed consent and with adverse outcomes for those individuals. The Beecher report of 1966 reviewed several unethical or potentially unethical studies funded by the government that caused or may have caused harm to participants (Beecher, 1966). Dr. Beecher conducted his review with a reported sense of urgency due to increased requirements at the time to test new treatments prior to implementation as well as due to the increasing availability of research funds providing more incentive to engage in such research. He also noted that physicians were under pressure to succeed as investigators when seeking promotion in academic settings.

The Beecher report noted an increasing presence of ethical errors, and Dr. Beecher addressed several types of these errors. He reviewed studies with ethical errors in various areas including: known effective treatment withheld (such as in the treatment of rheumatic fever); physiologic studies, such as of drugs or toxins with known deleterious effects, or of potentially harmful surgical procedures; studies to improve the understanding of disease; and a study using X-ray exposure in newborn infants to determine whether reflux occurs in the normal bladder. Dr. Beecher emphasized the need for *informed* consent as well as the need for a responsible and compassionate investigator to be present.

One study that gained some notoriety following its inclusion in the Beecher report was the so-called Willowbrook study. However, subsequent publications have suggested its inclusion was somewhat controversial (see Robinson and Unruh (2008), for a discussion of this and a review of the study). The study involved children with intellectual disability living at the Willowbrook State School. There, new infection (post admission) with a mild form of hepatitis was prevalent. Children were enrolled in a study of hepatitis and its prevention with the informed consent of their parents. In one phase of the study, children treated with gamma globulin were found to have protection from the hepatitis strain in comparison to an untreated comparison group (controls). A subsequent phase of the research was conducted to test the hypothesis that infecting individuals with the virus after injecting them with gamma globulin would induce passive–active immunity against the disease. The hope was that future cases of hepatitis at the institution (and elsewhere) could be reduced. In fact, there was reportedly an 80–85% reduction in incidence of hepatitis among the children and employees at Willowbrook.

As noted earlier, Beecher included the Willowbrook study as one of several unethical or questionably ethical studies. Notably, though, as reviewed by Robinson and Unruh (2008), the ethical issues involved had been carefully considered and methodically reviewed by the lead investigator, Dr. Saul Krugman, and his colleagues. For instance, informed consent had been obtained from the participants' parents, the study protocol and environment had been designed to present the least possible risk to participants, the investigators had considered the pre-existing risk to the children that was already present (i.e. the prevalence of the infection at the facility), they had carefully reviewed previous cases at the facility and learned that the cases were mild with no deaths, they had considered the potential benefits to the participants themselves in addition to other children facing the same pre-existing risks, and they had obtained an independent review of the study design from other experts in the field.

In spite of arguments made by various authors regarding the sound ethics of the Willowbrook study, concern about its ethics have continued to be raised as a point of contention in the conducting of research with vulnerable populations such as children and those with intellectual disability. Some authors caution, though, that concerns raised by studies such as the Willowbrook experiments may be over-applied and actually result in unnecessarily avoiding research among particular groups (e.g. women of childbearing age and children). This avoidance may lead to a lack of research with particular groups and, instead, an applying of research findings from one group to another when those original findings may not generalize across those groups. For instance, as pointed out by Robinson and Unruh (2008) in their review of the Willowbrook study and the fallout of the Beecher report, excluding women and children from research can and has led to assumptions about whether and how women (and children) benefit from the products of research in which those groups have been excluded due to ethical concerns. In other words, as pointed out by the authors, the unique medical issues of these populations are sometimes ignored with differential responses of the groups to standard care remaining unknown ("invisible"). The authors conclude in stating that it would be a "mistake to continue to allow the experiments at Willowbrook to cast a restrictive ethical pall over the participation of vulnerable children in medical research." The valid issues raised by Robinson and Unruh must, of course, be balanced against the risks to vulnerable parties. As part of that balance, the potential benefit of the research is of paramount importance in the moral and ethical equation.

In this discussion, I have reviewed instances of physical harm being intentionally or unintentionally perpetrated against research participants. Not all harm in research occurs in the physical realm, however. Researchers must also be aware of the potential risk of psychological harm that can occur as a result of research inquiries. Some of the potential psychological risks are reviewed next.

1.3 Ethical Issues in the Social Sciences

Different types of psychological harm can arise in a number of research contexts. In modern times, concerns have been raised about research in the social sciences involving deception. The use of deception is somewhat controversial, and studies of this type may leave participants psychologically affected.

One well-known study involving deception was the Milgram Experiment on obedience to authority figures (Milgram, 1963). In that study, participants were instructed to press a lever in order to administer shocks to another individual who was in a different room. The participant was told that the purpose of the study was to assess the impact of punishment on learning. In reality, the individual receiving "shocks" was a confederate who was not actually being harmed, but that was not known by the participants. The actual research participants who complied with the instructions were led to believe that they were actually inflicting suffering on another human being. Notably, they were pressured to do so. Based on Milgram's own description in his report, it is clear that a number of participants experienced distress during the actual experiment. Another concern to be considered, though, is the psychological burden that participants may have experienced upon realizing what they were capable of.

This latter issue is also a concern raised by opponents of bystander apathy studies. Many argue that studies of bystander apathy have the potential to leave participants feeling ashamed. At times, these studies may involve staged observational studies in which informed consent is not first obtained or studies in which deception is employed (i.e. participants do not know the true nature or purpose of the experiment). Some might even consider the popular ABC television show *What Would You Do?* (Quiñones, 2008–Present) to have the potential to harm unwitting participants upon learning that they had acted apathetically instead of intervening on the behalf of a confederate victim.

A good example of a bystander apathy study was an experiment by Darley and Latané (1968). Theirs was an experiment evaluating the phenomenon of diffusion of responsibility and its impact on people's willingness to intervene on behalf of other individuals. Participants in the study were told that they were taking part in a discussion about personal problems associated with college life, but that the conversations were being held via an intercom system to protect privacy. During the "discussion," a confederate faked a seizure. The researchers recorded the time it took for the participant to seek help for the "victim" in need. The purpose was to determine if that speed was influenced by the number of other "discussants" the participant believed could also hear the pleas of the victim. Shortly after each participant's responses were recorded, they were informed of the true nature of what had just occurred.

Empirical evidence for the potential psychological harm of bystander apathy studies and other research involving deception has been equivocal (Schwartz & Gottlieb, 1981). Regardless of whether evidence supports concerns about potential harm, though, there are steps that can be taken to reduce such harm and discomfort for the participants.

Pascual-Leone, Singh, and Scoboria (2010) review practical issues and evidence regarding ethical considerations when using deception. A primary factor addresses whether the benefits of the research outweigh the potential distress for the participants. When there is the potential for distress, researchers should provide debriefing and support following the experiment, and the respective IRB should evaluate the relative merits of the use of deception and whether debriefing should occur (Cheng-Tek Tai, 2012). There may be times when using deception is necessary and the research answers questions that may eventually improve understanding of human behavior in a way that has positive outcomes for society as a whole. In terms of debriefing, there may be instances in which the process may not be possible, including when there is the risk of subject pool contamination, when debriefing is not practicable, and when there is risk that the debriefing may be more harmful than the deception (Sommers & Miller, 2013).

So far, I have reviewed issues related to physical harm and psychological distress imposed on human research participants. Animals are also frequently utilized in research for a variety of reasons, and they can also suffer. In the following section, I review issues related to the welfare of animal subjects in research.

1.4 History of Harm to Animal Subjects in Research

Many of us can look back with fondness on memories of Elle Woods, J.D. fighting to save animals from unjust harm in research conducted solely for the purpose of developing cosmetics. While Ms. Woods was a fictional movie character with a propensity for all things pink, the issues highlighted in the movie were real. Animals can and do experience pain as has been demonstrated through scientific investigations, including studies showing that certain animals can be taught to avoid painful stimuli or even to seek pain-mitigating stimuli (for a review and discussion, see National Research Council (2009)). When used in research, then, animals may potentially suffer.

In fact, animals are frequently used in research, and this is due to several factors. These factors include their relatively short life cycle for evaluating effects of treatments over their lifespan, the ease of breeding and controlling genetic lineage, and the ease of reproducing large numbers of potential animal research subjects. Animals are also not free to decline participation in research making it even easier to acquire large sample sizes. Due to these factors and the potential for harm, there are now rules in place to protect the rights and welfare of animals (for instance, see National Institutes of Health (2015)).

In the story depicted in *Legally Blonde 2*, Ms. Woods addresses the rights of animals, and the story line focuses on the use of animals to evaluate cosmetics. In contrast to this scenario, much research is conducted using particular animals to study human diseases and response to treatments, as well as to develop vaccinations to prevent new cases of diseases. There are historical instances of lifesaving treatments emerging from such research, including the creation of vaccines against rabies, polio, diphtheria, and tetanus. Lifesaving antibiotics have also been developed thanks to animal research and testing, and organ transplant techniques have evolved with the aid of animal research. Drugs to treat various ailments have been developed, and surgical techniques have been improved through animal research. These and more medical advances are reviewed in a document published by the National Academy of Sciences (1991) along with a discussion of animal welfare issues and regulations to protect animal subjects in research.

Unfortunately, there have been incidents involving the inhumane treatment of animals, including in medical research. As noted by Franco (2013), examples of animal cruelty in the name of science date back several centuries with ancient Greek physicians conducting vivisections on animals for anatomical research as long ago as the sixth to fifth century BCE. As reviewed by Franco, studies of animal anatomy and physiology continued through the centuries, with the presence of competing philosophies regarding the sentience of animals.

In addition to potential physical pain to animals in the study of animal physiology, anatomy, and response to treatment, many individuals and groups have raised concerns about harm and suffering that occur in other types of research, including behavioral research. Groups have long objected to studies in which stress is imposed on animals to evaluate their responses in the study of depression and anxiety, for example. There have also been outcries against studies in which very young animals are separated from their mothers to assess the impact of separation and/or isolation. These are serious issues that researchers need to consider and address.

As noted earlier, there have been many benefits of having animals involved in research. In addition to the benefits for humans, animals do potentially benefit from the research, including that animals are sometimes treated with similar drugs, antibiotics, and vaccines as humans, and surgical techniques and developed pain-control measures can sometimes be applied to animals. However, some authors have noted that there are times when findings from animal research do not translate well to humans, and this may result in harm to humans and/or premature abandonment of potential treatments due to misleading findings from animal studies (Akhtar, 2015). Where there are benefits to employing animal testing, though, these benefits must still be balanced against the cost to the animals in terms of pain, suffering, and lack of freedom. Fortunately, there are steps that can be taken to minimize the suffering of animals, and there have been regulations developed to address this (to be discussed further in Section 1.6).

1.4.1   Summary

In the preceding pages, I have reviewed some dark history in the world of science and research involving humans as well as animals. In some of these cases, most of us would not consider the willful acts of malevolence to fall into the realm of research as they are more accurately characterized as torture and extreme cruelty. Some of the examples reviewed and others prevalent in our history, though, can be characterized as research with good intentions but sometimes ill results for the participants. It is because researchers can sometimes have good intentions while failing to be aware of or mindful of the needs of their participants that we need guidelines, rules, and regulations to protect potential victims. In the following pages, I summarize some of the major milestones in the evolution of research ethics, guidelines, and rules. In Table 1.1, I summarize the timeline of some of these milestones along with some of the more widely recognized occurrences of research harm.

1.5   Ethics, Principles, and Guidelines

As noted earlier, there have been debates about the humane treatment of animals and humans in research for centuries. In the mid-1800s and early 1900s, physicians and scientists were calling for the ethical consideration of research participants and even informed consent (see Jones, Grady, and Lederer (2016), for a discussion). Historical instances of unethical research, especially those during World War II, have led to the development of formalized codes of ethics, guidelines, and rules regulating scientists in the use of human participants and animal subjects. The over-riding goals of these dicta are to protect humans and animals from harm, and to ensure that there is benefit without undue suffering of participants or maleficence on the part of investigators.

The Nuremberg code stemmed from the Nuremberg trials, where individuals in the military as well as scientists were convicted of war crimes. These crimes included the atrocities perpetrated by the Nazis against prisoners in

Table 1.1 Timeline of events and the evolution of research ethics.

Year(s)	Event
Sixth to third centuries BCE	Vivisection and dissection of live animals are performed by ancient Greek scientists, with vivisection of convicted (human) criminals in the latter part of that period
1876	Britain's Cruelty to Animals Act – an extension of an earlier version of the Act – is established to protect animals from painful experimentation
1932–1972	Tuskegee Experiment
1935–1945	Unit 731 of the Japanese Imperial Army experiments on Chinese and Russian Prisoners of War; Experiments on American POWs at Kyushu University
≤1939–1945	Nazi Medical Experiments are perpetrated against innocent victims
1947	Nuremberg Trial occurs, and the Nuremberg Code is established
1951	Ethical Standards for Psychology, Section 4 – Ethical Standards in Research is prepared by the American Psychological Association Committee on Ethical Standards for Psychology (Cook et al., 1951)
1963	Milgram Experiment on obedience is reported
1964	World Medical Association adopts the Declaration of Helsinki
1966	The Beecher report is published
1966	American Medical Association adopts Ethics Guidelines for Clinical Investigation (American Medical Association, 1966a, b)
1966	The Animal Welfare Act is signed into law in the United States, creating animal welfare regulations (United States Department of Agriculture, 1966)
1974	National Research Act is signed into law in the United States, creating the National Commission for the Protection of Human Subjects of Biomedical and Behavioral Research
1979	The *Belmont Report: Ethical Principles and Guidelines for the Protection of Human Subjects of Research* is produced, and the National Commission for the Protection of Human Subjects of Biomedical and Behavioral Research is established
1981	United States Department of Health and Human Services signs a revised Code of Federal Regulations for the Protection of Human Subjects, which is later adopted in 1991 by 16 federal agencies involved in the conduct, support, or regulation of research involving human subjects; this group includes the National Institutes of Health (Rice, 2008)
1985	The Health Research Extension Act of 1985, Public Law 99–158, "Animals and Research" is established as law mandating the development of Public Health Service Policy on Humane Care and Use of Laboratory Animals
1994	NIH Guidelines on the Inclusion of Women and Minorities as Subjects in Clinical Research are established
2009	Presidential Commission for the Study of Bioethical Issues is created

their concentration camps (Fischer, 2006). The code was a culmination of evolving views regarding ethical research practices, and summarized as "Permissible Medical Experiments" in the documentation of the Nuremberg Military Tribunals (Germany {Territory under Allied occupations, 1945–1955; U.S. Zone}, 1949). Codified was the essentiality of voluntary consent of human participants ("subjects"). Other aspects of the code included the justification of the research based on potential outcome versus human cost, minimization of harm or injury, the ability of the participants to withdraw from the research, and quality assurance.

In 1964, the World Medical Association adopted the Declaration of Helsinki. The Declaration outlined principles for ethical research, and has been amended several times (World Medical Association, 1964–2013). Many of the principles parallel the Nuremberg code, but with some additions. Among the additions is the admonition that under-represented groups should be given access to research participation. Another principle states that individuals harmed by research must be compensated and treated.

Following the harrowing revelations of the Tuskegee experiment, the US government took steps to prevent future such atrocities. The National Research Act was signed into law in 1974. That Act included the creation of the National Commission for the Protection of Human Subjects of Biomedical and Behavioral Research (see Centers for Disease Control and Prevention (2017a), for a review). That panel's recommendations have evolved through several iterations over the years, most recently with the Presidential Commission for the Study of Bioethical Issues in 2009 (see US Department of Health and Human Services, 2017). The Commission's purpose is to encourage policies and practices ensuring that research and other health-care activities are conducted in a socially and ethically responsible manner. The National Commission has worked to achieve this through the completion of projects and the publication of reports addressing a range of relevant topics.

For instance, the National Commission for the Protection of Human Subjects of Biomedical and Behavioral Research (1979) generated the *Belmont Report*. The *Belmont Report* includes a list of ethical principles and guidelines for the protection of human subjects of research. The report distinguishes between practice and research, and clarifies the need to make that distinction in order to know when and what activities should undergo review for the protection of research subjects. The authors of the *Belmont Report* additionally review basic ethical principles, including: (1) Respect for Persons, (2) Beneficence, and (3) Justice.

Respect for persons means that individuals must be treated as autonomous, and persons with diminished capacity are entitled to protection. Beneficence refers to the rules to do no harm, and to maximize possible benefits and minimize possible harms. Justice relates to the manner in which subjects of research are selected such that people are not selected on the basis of easy availability, compromised position, or manipulability as opposed to reasons directly related to the topic under investigation. Justice further relates to the manner in which the benefits of the research are ultimately distributed, in particular, by not excluding particular groups from benefit, and not involving individuals in research if they will likely not benefit from subsequent applications of the research. Finally, the *Belmont Report* lists a set of Applications stemming from those three basic principles. The Applications include: (1) Informed Consent, (2) Assessment of Risks and Benefits, and (3) Selection of Subjects.

Not to be ignored, animal welfare has also been addressed through the development of guidelines for the humane treatment of animals in research. The Health Research Extension Act of 1985, Public Law 99–158, "Animals and Research" was established as law mandating the development of new policy. That policy, Public Health Service Policy on Humane Care and Use of Laboratory Animals, has been updated and revised over the years, and includes guidelines for the care and treatment of animals in research (National Institutes of Health Office of Laboratory Animal Welfare, 2015).

Professional organizations governing scientists have also developed ethical research principles. Among those are the American Medical Association (AMA, 2018) and the American Psychological Association (APA, 2016). The principles of these organizations serve as guidelines and aim to protect research participants. Again, many of the outlined principles parallel those of the Nuremberg code and the Helsinki Declaration. Inclusive, both sets of guidelines highlight the importance of informed consent.

The American Psychological Association's research ethics were initially prepared to be included in their already existing general ethical guidelines and published in tentative form in 1951 with an invitation for members to consider the guidelines and provide feedback (Cook et al., 1951). In these guidelines, the APA Committee on Ethical Standards for Psychology addressed the need to plan research in such a way as to avoid the possibility that the findings will be misleading. Inclusive, they discussed selection bias, omission of control groups, and lack of generalizability. They also addressed the need to report findings completely and accurately, and to not withhold negative findings. Most importantly, in the opinion of many, the guidelines addressed the protection of subjects. Inclusive, they discussed the need for informed consent with some exceptions, as well as the need for confidentiality. Finally, the guidelines even addressed the obligation to follow through with the promise of rewards or services made in exchange for participation in the research. The ethical guidelines of the APA have continued to be expanded, revised, and updated over the years.

The American Medical Association (AMA) published their initial set of research ethics guidelines under the title *Ethical Guidelines for Clinical Investigation* in 1966. The House Delegates of the AMA endorsed the ethical principles of the 1964 Declaration of Helsinki of the World Medical Association noting that they conformed to the fundamental concepts incorporated into the AMA's *Principles of Medical Ethics* (American Medical Association, 1966a, b). In addition to informed consent and beneficence, these principles addressed the need for physicians to recognize the existence of the physician–patient relationship and the need to exercise his (or her) judgment in the best interest of the patient. They also addressed welfare, safety, and comfort, as well as special protections for minors, and the need for voluntary versus forced participation. As with the American Psychological Association's code of ethics, the AMA's ethical guidelines have continued to be expanded, revised, and updated over the years.

In 1994, the National Institutes of Health (NIH) addressed the issue discussed earlier in this chapter regarding the need to include women of child-bearing age in research. The NIH published the *NIH Guidelines on the Inclusion of Women and Minorities as Subjects in Clinical Research* (1994). These guidelines were intended to strengthen previous NIH policy documents from 1990. According to the authors of these guidelines, the policy "should result in a variety of new research opportunities to address significant gaps in knowledge about health problems that affect women and racial/ethnic minorities and their subpopulations." The spirit of the guidelines can be thought of as encouraging and facilitating the inclusion of groups as research participants in order to evaluate the differential effects of treatments and conditions and to avoid inappropriate generalization of findings from one group to another. These guidelines have subsequently been updated and revised, most recently in 2017 (National Institutes of Health, 2017).

1.6 Statutes and Regulations Protecting Humans and Animals in Research

While the above principles serve as guidelines, scientists' engagement in ethical research practices is typically regulated by statute as well. For example, psychologists licensed in Texas where I practice are regulated by the Texas State Board of Examiners of Psychologists (TSBEP). The TSBEP develops and enforces rules of practice, including rules addressing research with humans and animals. Notably, the TSBEP rules pertaining to the conduct of research parallel the American Psychological Association's Ethical Guidelines. When embarking on research, scientists should familiarize themselves with the relevant statutes and regulations in their jurisdiction.

Federal regulations for the protection of human research participants in the United States are covered in the Code of Federal Regulations, Title 45, Subtitle A, Subchapter A, Part 46 – Protection of Human Subjects (Government Publishing Office, 2018). The Secretary of the Department of Health and Human Services signed a revised Code of Federal Regulations for the Protection of Human Subjects back in 1981. Those regulations were adopted in 1991 by 16 federal agencies involved in the conduct, support, or regulation of research involving human subjects, including the NIH (Rice, 2008). The policies outlined in the code apply to government employees and institutions as well as to research supported by the federal government. The code addresses what types of research require institutional review, and it outlines the composition and role of the IRB (which I will be discussing throughout the remainder of this chapter). Informed consent is covered extensively, and there are policies specifically addressing protections for pregnant women, children, human fetuses, and neonates involved in research. Regulations for the protection of prisoners involved as research participants are additionally incorporated into the Code.

Also at the federal level, regulations have been promulgated to protect animals when used in research and testing. In 1966, the Animal Welfare Act was signed into law in the United States. The Animal Welfare Act has been revised and updated numerous times over the years. The Act and its revisions call for and create regulations mandating procedures striving for the humane care and handling of animals. The current regulations are found in the United States Code of Federal Regulations, Title 9, Subchapter A-Animal Welfare (Government Publishing Office [US], 2016). Among other stipulations, the regulations require regular inspection and guidance from animal experts as defined by the regulations, as well as the potential for suspension of activities involving animals when there are violations of the regulations.

Regulations protecting research participants are not unique to the United States. The United States Department of Health & Human Services Office for Human Research Protections has compiled a list of more than 1000 laws, regulations, and guidelines from 130 countries and international organizations (US Department of Health & Human Services, 2016). The so-titled *International Compilation of Human Research Standards* is composed of laws, regulations, and guidelines categorized into nine groups: (1) General; (2) Drugs and Devices; (3) Clinical Trial Registries; (4) Research Injury; (5) Social-Behavioral Research; (6) Privacy/Data Protection; (7) Human Biological Materials; (8) Genetic; and (9) Embryos, Stem Cells, and Cloning.

Predating all of the above guidelines and regulations, Britain extended its 1835 Cruelty to Animals Act in 1876 (Cruelty to Animals [39 & 40 VICT. Ch. 77.], 1876). The extension included law prohibiting painful experimentation on animals. Also prohibited were the use of urari (related to curare) as an anesthetic in animal testing and public exhibition of painful experimentation.

At the local level, most institutions, including universities and medical schools, have agencies responsible for oversight of research by their personnel, students, and trainees. These are typically referred to as the IRB and/or Institutional Research Committee (IRC). The IRB/IRC is responsible for reviewing, approving, and monitoring research. An important mission of the board or committee is to ensure the safety and wellbeing of human research participants and animal subjects. Like state rules, many of the requirements of the IRB incorporate the ethical research principles that have been discussed throughout this section.

1.7 More About Informed Consent

One of the most important concepts covered in ethical guidelines as well as statutes is that of informed consent. In this context, the term "informed consent" refers to the need for participants in an activity, whether research or treatment-related, to be made aware of all the facts, potential and real, related to their participation. It is the right of every human being to know what will occur, any risks, potential benefits, fiscal costs or incentives, time requirements, alternative options, how and where to access information or to obtain help in regard to participation now and in the future, and their choice in terms of consenting or not. There may be instances in which an agency or review board determines that observational or archival studies do not pose any risks or potential for harm, and thus participants may not be informed or asked to consent to participation. In those instances, though, great care must be taken to avoid unanticipated harm, and there should be continued reassessment of that potential.

If an agency such as an IRB has determined the potential for harm or the need for individuals to be informed, then informed consent must be obtained from participants. Informed consent is almost always obtained in writing. Note that the concept of "informed consent" goes two ways – that is, it involves both providing information and obtaining consent, or agreement to participate. That agreement is voluntary, a point that must be made clear to the potential participant.

Children represent a special consideration with respect to informed consent. Legally, in most, but not all, jurisdictions, children cannot give consent before the age of 18. Again, this depends on jurisdictions as well as whether the child has been legally emancipated. In most instances, the parent or legal guardian must provide consent before a minor can participate in a research study. However, regardless of age, children can additionally give "assent." Some argue, with strong evidence, that children have the capacity to give consent at a much younger age than 18 (Hein et al., 2015), which makes it even more important to provide full information and obtain, at the least, assent from the minor. Your IRB/IRC will also have requirements for obtaining consent and/or assent depending on the age of the child.

In Table 1.2, I include a sample Consent Form. This consent form is based on a hypothetical (a.k.a. fictional) research study. There are many consent form templates available online, with various formats addressing different types of research (e.g. behavioral studies, clinical trials). If you are part of an institution (e.g. university or medical school), it is likely that your IRB will have templates available to you.

Table 1.2 Sample informed consent form.

State University Human Research Protection Department
Informed Consent Document

Project Title: Sleep, Exercise, Depression, and Cognition

Institutional Review Board Number: 1234567

Approval Date: December 31, 1999

Expiration Date: January 2, 2009

You are invited to take part in a research study being conducted by Dr. Pat Finkleberry, a researcher at State University, and funded by State University Medical School. The information in this form is provided to help you decide whether or not to take part. If you decide to take part in the study, you will be asked to sign this consent form. If you decide you do not want to participate, there will be no penalty to you, and you will not lose any benefits you normally would have. You may choose to withdraw from the study at any time without penalty.

NOTE: If you are employed, then it is your responsibility to work with your employer regarding work leave for participation in this study if during work hours.

Contact Information: If you have any questions during or after this study, you may contact Dr. Finkleberry at this office by phone at 555-555-5555 or by e-mail at Pat. Finkleberry@email.com.

Study Purpose:

The purpose of this study is to evaluate the effectiveness of certain types of educational instruction in managing depression and improving cognitive skills. We hope to use this information to help other patients in the future.

Why Have I Been Asked To Be In This Study?

You are being asked to participate in this study because you are a patient in our Neuropsychology Clinic, and we are interested in the helpfulness of our educational strategies for people referred for neuropsychological evaluations. The neuropsychologist has already determined that you meet the criteria to be in this study.

How Many People Will Be Included In This Study?

136 people (participants) will be invited to participate in this study from this clinic.

Are their Alternatives to being in this Study?

The alternative to being in the study is to not participate. You will continue to receive the services for which you were originally referred. The neuropsychologist provided his/her treatment recommendations during the feedback session you just had. The research that you are being recruited for is separate from, and in addition to our standard treatment.

What Will I Need To Do In This Study?

You will be randomly assigned to one of four groups. Depending on which group you are assigned to, you may be asked to participate in a one-hour session in which you will receive education about activities and strategies that may improve your mood and thinking. You may also be given written materials to take home with you. Everyone in the study will be scheduled for a follow-up appointment in six months; we will work around your schedule when the appointment is made. During that appointment, you will be interviewed about your activities during the six-month period, and you will complete one questionnaire and one cognitive test; that appointment should last less than one hour. A research assistant

(Continued)

Table 1.2 (Continued)

will obtain information from your neuropsychological clinic record to combine with the new information that we obtain from you during the follow-up appointment.

_____ I give my permission for my neuropsychological evaluation records from this clinic to be accessed for use in this research study.

_____ I do not give my permission for my neuropsychological evaluation records from this clinic to be accessed for use in this research study.

Are There Any Risks Involved in My Participation?

The things that you will be doing are no more than the risks that you would come across in everyday life. You will be answering a questionnaire similar to one you completed today as well as at least one cognitive test that you took today. Although the researchers have tried to avoid risks, you may feel that some questions that are asked of you will be stressful or upsetting. You do not have to answer anything you do not want to.

Are There Any Benefits To Participating In This Study?

The direct benefit to you by being in this study is that you may experience improvement in your mood or thinking, though we cannot promise that since this intervention is experimental.

Will There Be Any Costs To Me?

Aside from your time, there are no costs for taking part in the study.

Will I Be Paid To Be In This Study?

You will not be paid for being in this study.

Will Information From This Study Be Kept Private?

The records of this study will be kept private. No identifiers linking you to this study will be included in any sort of report that might be published. Research records will be stored securely and only Dr. Finkleberry, the research assistant, and authorized research personnel described below will have access to the records. Information about you will be stored in a locked filing cabinet and in computer files protected with a password. This consent form will be filed securely in an official area as well.

People who have access to your information include the Principal Investigator (Dr. Finkleberry) and the research assistant. Representatives of regulatory agencies such as the Office of Human Research Department and entities such as the State University Human Research Protection Department may access your records to make sure the study is being run correctly and that information is collected properly.

Research personnel at State University Medical School, which is funding this study, may also see your information. However, any information that is sent to them will be coded with a number so that they cannot tell who you are. Representatives from the research review board can see information that has your name on it if they come to the clinic to view records. If there are any reports about this study, your name will not be included in them.

Information about you related to this study will be kept confidential to the extent permitted or required by law.

Incidental Findings: If the research assistant learns of any concerning issues during your follow-up appointment, she will notify Dr. Finkleberry immediately so that Dr. Finkleberry can discuss this with you as well as any recommendations for intervention. You and Dr. Finkleberry will decide together whether you should be referred to your doctor or other provider.

Table 1.2 (Continued)

Whom May I Contact for More Information?

You may contact the Principal Investigator, Pat Finkleberry, Ph.D., to report a concern or complaint about this research at 555-555-5555 or Pat.Finkleberry@email.com. You may also contact the research assistant, Bugs B. Unny at 555-555-5555 or BBunny@email.com.

For questions about your rights as a research participant, to provide input regarding research, or if you have questions, complaints, or concerns about the research, you may call the State University Human Research Protection Department by phone at 1-555-555-5555, toll free at 1-800-555-5555, or by e-mail at irb@su.eddu. The informed consent form and all study materials should include the IRB number, approval date, and expiration date. Please contact the Human Research Protection Department if they do not.

Am I Free To Change My Mind About Participating?

Your participation in this research is voluntary, and you have the choice whether or not to be in this research study. You may decide to not begin or to stop participating at any time. If you choose not to be in this study or to stop being in the study, there will be no effect on your care or treatment in this clinic or in your relationship with State University Medical School. Any new information discovered about the research will be provided to you. This information could affect your willingness to continue your participation.

STATEMENT OF CONSENT

I agree to be in this study and know that I am not giving up any legal rights by signing this form. The procedures, risks, and benefits have been explained to me, and my questions have been answered. I know that new information about this research study will be provided to me as it becomes available and that the researcher will tell me if I must be removed from the study. I can ask more questions if I want, and I can still receive services if I stop participating in this study. A copy of this entire consent form will be given to me.

_____   _____
Participant's Signature   Date

_____   _____
Printed Name   Date

INVESTIGATOR'S AFFIDAVIT:

Either I have or my research assistant has carefully explained to the participant the nature of the above project. I hereby certify that to the best of my knowledge the person who signed this consent form was informed of the nature, demands, benefits, and risks involved in his/her participation.

_____   _____
Signature of Presenter   Date

_____   _____
Printed Name   Date

For the sample consent form in Table 1.2, I borrowed heavily from a template available online through Texas A&M University (2012), which houses medical and veterinary schools as well as several academic undergraduate and graduate schools. In this example, the study being consented to involves very minimal risks, and there is no compensation. Most IRBs will require a thorough review of risks, compensation and/or treatment for those risks (or lack thereof), and any payment to be received or paid by the participant. Participants should also be informed of services available if they should suffer an injury or other adverse event as a result of being in a study that has potential risks. There may even be mandated language for this if the study is sponsored by a pharmaceutical company.

1.8   The Importance of Freedom to Withdraw

As part of the informed consent process, research participants must be notified from the start that they have the right to withdraw from the research at any time. Astonishingly, this concept is sometimes lost on researchers who believe that a commitment to participate in research is equivalent to a contract and requires the participants to see things through to the end. This is just not so, and it is imperative that the researcher is aware of that, and that the participant is informed of the right to withdraw from the beginning and at any time going forward. It must also be made clear to the participant that there will be no adverse consequence if they choose to withdraw. The clarification of the right to withdraw from research is particularly critical in the clinical setting where the researcher may also be, or may be associated with, the potential participant's health-care provider.

1.9   Separation of Provider–Researcher Role

Closely related to the issue of undue influence (to be discussed next) is the need to separate the role of the investigator from the role of the clinician and treatment provider. While undue influence addresses the need to avoid pressuring patients to participate in a research study, separation of the provider–researcher role addresses the need to provide good care for your patient without being influenced by the desire to recruit more participants. That is, the needs of your patient as a patient should always come before the needs of your research.

Think about cancer patients recruited for clinical trials. They may confuse the recruitment process with their provider's treatment recommendations. Or, they may feel torn between their drive for the best chance of survival and their desire to please their health-care provider on whom they have come to rely. It is important for the researcher to be aware of their sometimes necessary, but at the same time complicated, dual role. They are conducting research, but they are also the person that the participant patient is looking to for advice and guidance on treatment options.

Above all, the clinical researcher must put their role as health-care provider to the patient first. Secondly, and equally critical, the clinician must inform and reassure the patient participant that she or he has the freedom to withdraw from the research without losing their clinical care from the provider.

To prevent a blurring of the investigator–provider role, it is ideal if participant recruiting/enrollment is completed by a research assistant or other clinician if there is any risk to the patient's care. One way to assess this risk is to ask yourself if your ability to recommend the best treatment option to your patient is in any way impacted by your need to recruit and/or randomize participating patients into one treatment arm or another. If there are other clinical trials or approved treatments that may offer a better outcome for your patient, you may not be unbiased in your recommendations if you are recruiting for your own study or trial. To make this less abstract, I have written a mock recruiting scenario in Table 1.3.

Table 1.3 Example of poor separation of investigator–clinician role.

Mr. Doe has recently been diagnosed with a recurrence of a rare disease. The long-term survival rate with current standard treatment is 40%. His specialist physician, Dr. Aspiring, is recruiting participants for a clinical trial for a new treatment method he is evaluating. The treatment is beginning Phase III trials. There is some indication that the new treatment may improve long-term survival slightly, but long-term side effects are not yet known as the Phase II clinical trials were just completed earlier this year. Also, because Dr. Aspiring's new treatment is in Phase III trials, Mr. Doe could be randomized to the old, standard treatment, and he will not know until after the study is terminated which treatment he received. There are three other new treatments currently being tested in Phase II and III trials, and one of those with promising results may receive FDA approval in the next six months or so.

Mr. Doe looks to Dr. Aspiring, his physician, for advice on which treatment he should opt for. However, Dr. Aspiring is conducting his own recruiting for his trial, and asks Mr. Doe if he will consent to participate. When Mr. Doe asks Dr. Aspiring which treatment he recommends, or whether Dr. Aspiring thinks Mr. Doe should hold out for the other treatment that may receive FDA approval soon, he does not get a straight answer. Dr. Aspiring tells Mr. Doe that he must decide for himself, and that once he decides, it is important for him to remain in the study for two years so as not to disrupt the integrity of the randomization. Also, he could not possibly have both treatments due to potential drug interactions and toxicity.

Mr. Doe does not know what to do. He does not know what the best option is for his long-term survival, so he seeks a second opinion. Fortunately, he is aware that he can seek another opinion, and he has the resources to do that. He feels somewhat dejected at his inability to get Dr. Aspiring's opinion, and he also feels like Dr. Aspiring is annoyed with him for asking. His doctor–patient relationship and confidence in Dr. Aspiring has been broken.

Note: In addition to disrupting the doctor–patient relationship due to blurring of the investigator–provider roles, Dr. Aspiring also violated a major tenet of research ethics; that is, he suggested that Mr. Doe should not withdraw from the study once he is enrolled. Participants must be informed that they are free to withdraw from participation at any time and without negative consequences. Otherwise, they will have been coerced to remain in the study.

1.10 Undue Influence

If the above two components of the consenting process (duty as clinician and the right to decline or withdraw) are not conveyed clearly and strongly to the potential research participant, then there is the potential for undue influence. That is, patients may agree to participate and/or remain in a research study either against their own judgment or, at the very least, with a degree of discomfort and lack of desire and true consent.

To minimize the risk of undue influence, the provider must take great care in conveying to their patient that their option to decline participation in the study will not influence the care they receive from the provider. They must know that the care will continue as before, and with 100% dedication to their treatment.

An alternative solution to avoiding undue influence is to recruit participants from a pool of patients not under the direct care of the researcher. This can be difficult in a setting in which patients are being treated for rare disorders, in which specific providers are known specialists in the rare disease. Those providers are also likely to be the ones conducting research involving new treatments. Even when patients are treated by someone other than the researcher or principal investigator, there is still likely a real or at least perceived association between the provider and the researcher when they are within the same treatment facility or institution. Therefore, the possibility of undue influence cannot be removed entirely.

1.11 Anonymity

Another critical aspect of research ethics is that of anonymity. Anonymity and confidentiality are important in protecting participants from stigma or other forms of potential harm. Additionally, promising and providing anonymity to potential research participants will likely improve recruiting for any given study. Every time I review my informed consent for treatment form with patients, I include a statement (orally and in the written form) that their test scores may be used in a research study, but that their name or other identifying information will not be included in any research reports. I often see on patients' faces a brief moment of apprehension when I mention research, followed by relief as I finish my sentence with the promise of anonymity. If I did not provide that information about anonymity, I suspect many would either balk at the idea, or feel unduly influenced to proceed in spite of their concerns about confidentiality.

Promising and providing anonymity and confidentiality requires a conscious effort to protect identifying information. This effort does not only apply to the dissemination and publication of research findings but also to the acquisition and storage of research data and records. The issue of anonymity in data acquisition conveys to the concept of privacy. That is, participants' privacy should be respected, and access to their data should not be given freely. Only individuals who are officially approved by the IRC or IRB as part of the research personnel should have access to participant data. Those personnel, in turn, must be trained and credentialed in issues related to privacy, including with respect to relevant statutes and institutional policies.

Directly related to the concept of anonymity is the need for secure storage of research data. Data, including raw or hard copies of research data and materials, must be kept secure in a manner and location that prevents others from accessing the data with associated identifiers. Again, all research personnel with access to or responsibility for gathering data should be approved and fully trained and credentialed in the requirements and methods for storage and security.

IRBs will typically have very specific requirements for maintaining research data in a manner that ensures anonymity. At the same time, in certain types of research (e.g. drug trials), there will also be procedures in place for coding patient data such that the patients can ultimately be identified by the proper personnel when or if there is a need to determine the arm of treatment (e.g. which drug) the participant received. This may be necessary if the drug is found to be beneficial so that those who may have received the placebo or less effective treatment can have the option of then receiving the new drug. Additionally, if a patient participant experiences a reaction or other adverse event, the clinicians involved in caring for the patient will need access to the patient's treatment history. So, the actual research data contained in a data set may not have the direct patient identifiers (e.g. name, date of birth, or medical record number), but there will be a research identification code. That identification code will be associated with the actual patient identifiers in a separate data set or list such that each patient in the list can ultimately be paired with his or her outcome and treatment information.

1.12 Summary

Research is an important tool for learning, increasing knowledge, and developing new procedures. In science, research often leads to improved methods of caring for groups and individuals, which may extend life or improve quality of life. At the same time, as outlined earlier in this chapter, there is great risk of harm when research is conducted in an unethical manner. In some instances, the harm is egregious and the tyranny in the methods is obvious. More often, the flaws that lead to harm are less obvious, even to the perpetrator. It is, thus, critically important for individuals who embark on research to be educated about the potential risks involved in research as well as the ethical guidelines, rules, and local requirements for conducting research. This is especially critical when engaging in research involving human participants and animal subjects, though research ethics and issues also apply to scientific investigations involving the environment or any matter that may ultimately impact our existence.

Section 2

Basic Research Designs and Validity

In this section, I review issues related to research validity and research design. Research validity is a concept that relates to the ability to draw conclusions based on research findings and to apply those findings outside of the laboratory setting. Research design refers to the manner in which research is conducted, including with respect to recruitment and assignment of participants, administration of experimental procedures (if any), and data collection, analysis, and interpretation. A study's design directly influences its validity.

The next two chapters are intended to aid in the understanding of research validity and design. An understanding of these topics is important for designing valid studies as well as for judging and interpreting others' published research findings. Without a sound and valid study design, research findings are not meaningful. Invalid research conclusions can even be misused, and they can be frankly harmful. A researcher may employ the most technical statistics with the largest sample size imaginable, but ultimately the results will only be as good as the research design and the validity of their study.

Research Design and Analysis: A Primer for the Non-Statistician, First Edition. Leslie D. Rosenstein.
© 2019 John Wiley & Sons, Inc. Published 2019 by John Wiley & Sons, Inc.

2

Research Validity

Research often involves comparing one group to one or more other groups. In this type of research, we may be seeking to answer broad questions about group differences or we could be evaluating the impact of an intervention on the outcome. For instance, we can conduct a study to compare improvement in depression among individuals receiving a new treatment intervention versus individuals receiving the standard care. Similarly, we can compare two groups, say individuals with learning disorders and those without, on a measure of processing speed in order to learn if learning disorders are associated with slow processing speed.

We can also assess the relationship between two or more variables. This may tell us whether certain characteristics in the population are related to other variables, such as the outcome in certain disease conditions. For example, we may conduct a study to determine whether there is a relationship between the duration of disease in patients with Type 2 diabetes and brain atrophy; a positive correlation between disease duration and atrophy would suggest an association between the disease duration and increased atrophy. Additionally, we can assess the relationship between characteristics, such as foot size and height, with a significant positive correlation indicating that bigger feet are associated with taller height and vice versa.

Typically, when we conduct a study, it is to answer a specific question for a specific purpose. However, our ability to draw conclusions based on our research findings will depend on the validity of the study in terms of its design, the sample studied, and how cleanly it is conducted. Whether we can conclude one event led to another, or whether a study's findings generalize beyond our participants at this time and place will depend on several factors. These factors can be characterized as falling into one of two basic areas: internal and external validity.

In the next section, I review the concept of *internal validity* in more detail. I follow that with a review of *external validity* in Section 2.2.

Research Design and Analysis: A Primer for the Non-Statistician, First Edition. Leslie D. Rosenstein.
© 2019 John Wiley & Sons, Inc. Published 2019 by John Wiley & Sons, Inc.

2.1 Internal Validity

Internal validity affects the researcher's ability to deduce that "A" led to "B" (A → B). In other words, without internal validity, researchers cannot deduce causality; they cannot conclude that their intervention or other independent variable caused a change in their outcome or dependent variable.

One strategy to improve the odds of being able to conclude that there is a causal link between variables is to use an experimental design. A true experiment is one in which the independent variable (e.g. the treatment arm, the condition) is manipulated or created, participants are randomly assigned to levels of the independent variable, and extraneous influences are controlled and equal at all levels of the independent variable. Manipulating or creating the independent variable means that the researcher has created and/or chosen the condition to which a participant is assigned, such as an intervention or a control group condition. Examples of independent variables that *cannot* be manipulated by the researcher are age, place of birth, and hand preference. Random assignment indicates that an unbiased method of assigning participants to groups (e.g. intervention group or control group) was used in order to prevent pre-existing differences between the groups that could otherwise confound research findings.

Unfortunately, it is not always feasible to conduct a true experiment, but that does not make research useless. However, again, causality cannot be concluded when a study is not experimental by design.

The purpose of random assignment, manipulation of the independent variable, and control of extraneous variables is to maximize internal validity. Common threats to internal validity that are managed with these strategies include: history, maturation, measurement error, bias in participant selection and assignment, experimental mortality/attrition, experimenter bias, expectation, and sensitization caused by pretesting. Other authors may categorize these threats slightly differently, and additional threats should be considered as well (e.g. design contamination caused by experimental groups influencing one another). Here, I will focus on the major threats listed above.

2.1.1 History

History is a threat that cannot be controlled, but it can be managed. History refers to events that occur over the course of time during which an experiment is being conducted. For instance, if a researcher is studying the impact of an intervention, say a six-month course of therapy to treat depression, events outside of the experiment may occur and affect the targeted symptoms. As a fun, but realistic hypothetical example, researchers in Houston may have wanted to determine whether their new therapeutic intervention improves symptoms of depression, but they conducted their study in the late summer/early fall of 2017. Coincidentally, in the fall of 2017, the Houston Astros beat the Los Angeles

Dodgers in the World Series. So, if the participants in the study showed improved mood, how did the researcher know whether this was due to their new intervention, or simply because the participants were happy their home team won the World Series? This is a blatant hypothetical example of how history may impact the internal validity of a study and affect whether a researcher can conclude A → B, that is, causality. History can also work the other way, and diminish what otherwise could have been a true causal effect. In the same example from Houston, a negative historical event could have occurred (e.g. war, natural disaster, a major tragedy), increasing depression *in spite of* the treatment intervention.

As noted above, it is not possible to control history, but the threat that history has to an experiment's validity can be managed. How does one manage history in the context of research? By employing a control group. That is, instead of simply evaluating change over time in one group that receives the intervention, for example, the researcher can include a second group that does not receive the intervention. Both groups will experience roughly the same history, but only one will receive the treatment. Thus, the researcher can have increased confidence that any change in the intervention group relative to the control group is due to the treatment, and not some historical event. Control groups can include individuals assigned to a wait list, a group receiving standard care, or a placebo-control group. I will talk more about placebo-control groups when I review participant expectation.

2.1.2 Maturation

Another, related, threat to internal validity is maturation. Maturation is easiest to conceptualize in the context of pediatric or geriatric research. Maturation occurs naturally over time, with either an upward or a downward trajectory. Size, for instance, tends to have an upward trajectory over time in pediatric populations, while having a downward trajectory in older populations. So, if size is the dependent, or outcome, variable in a research study, some changes may occur solely due to maturation, and not due to the independent variable or intervention.

Maturation is not always tied to development or lifespan changes. For instance, many diseases or conditions have a natural tendency to improve or worsen over time. Progressive dementias, as one example, worsen over time by definition. Depression often improves spontaneously over time. Many viral illnesses also improve over time as long as there are no complications.

Like history, maturation cannot be controlled, but it can be managed with the use of a control group. Theoretically, a control group will undergo approximately the same natural changes over time as the experimental group, but with the experimental group additionally experiencing change due, for instance, to the intervention.

2.1.3   Measurement Error

One of the most important threats to internal validity is measurement error. Addressing measurement error is critical not only in ensuring internal validity but also in maintaining power. Power in research is the ability to detect an effect – this concept is reviewed in depth in Chapter 4. Measurement error is anything that reduces the reliability or precision of the assessment or quantification of the dependent (outcome) variable. Measurement error can occur as a result of poor calibration of an instrument, human error (even with data entry), and lack of test reliability (repeatability) and validity.

Reliability is the degree to which a tool obtains the same results across repeated measurements (test–retest reliability) and with different agents (inter-rater reliability). There are other variations of reliability, but for the purpose of this discussion, test–retest and inter-rater reliability are most germane. Lack of reliability can result in a phenomenon called *regression to the mean*. In regression to the mean, extreme scores will become less extreme, closer to the mean, on repeat assessment. Thus, a change in the score in one direction or the other may not have anything to do with the independent variable/intervention but, rather, may be simply the result of this statistical phenomenon. Lack of reliability creates noise, or error variance, that is not related to the construct being tested or the intervention being tested.

Related to reliability is the concept of *test validity*. Test validity is the extent to which the quantification of the dependent, outcome, variable actually measures the intended construct. There are various types of validity, including construct validity, content validity, predictive validity, and face validity. Face validity is not actually something a researcher necessarily strives for. Face validity is the degree to which a test appears to measure what is intended. In some cases, a researcher may not actually want the participants to know what the test is measuring because it may bias their responses. That concept will be discussed further when the impact of expectation is discussed. Construct, content, and predictive validity each have different merits, depending on the goal of the study. The important takeaway is that a study's results will be uninterpretable or meaningless if the measure used to assess the dependent, outcome variable is not actually measuring the intended construct. There are many good resources on the Internet and in textbooks for more information on test validity, with two old classics being *Psychometric Theory* (Nunnally, 1994) and *Psychological Testing* (Anastasi & Urbina, 1997).

The best way to manage measurement error is to eliminate it to the extent possible. This can be accomplished by utilizing precise measurement tools and ensuring quality assessment. This might involve having two independent parties measure the dependent variable when possible. This can be accomplished with variables such as height, weight, blood pressure, etc. in which issues such as practice effects are not relevant. Related, tools used for measuring the dependent variable should be checked regularly to ensure proper and consistent calibration.

When measurement tools are not concrete, it is important to use measures with proven good reliability and validity. Validity and reliability of a measurement instrument should be available from the test publisher as well as through independent research available in the literature.

Data entry can be checked and rechecked for accuracy by research personnel. When it is not possible to check all data due to the enormity of the data set, at least a subset of the data should be checked for data entry errors by randomly selecting a predetermined number of cases to be checked.

To the extent that measurement error cannot be eliminated, it can sometimes be managed, again, by including at least two groups in the study (e.g. an intervention group and a control group). Theoretically, each group will be impacted approximately equally by the measurement error, so any differences between the groups can be attributed to group differences (e.g. the intervention) and not the measurement error. However, if the measurement error is somehow biased, or different across groups, it will alter the extent to which causality can be deduced.

One issue with the measurement error that cannot be managed with the use of a control group relates to how variables are defined. In particular, this is an issue when considering diagnostic groups. If diagnoses are not made in a consistent manner using preset criteria, the findings of the study will be flawed, and potentially overstated. This problem arises especially with the use of databases, when electronic medical records are culled for the presence of diagnoses among patients. Those diagnoses may not have been made consistently, using the same set of criteria, across patients. Sometimes, diagnoses included in electronic medical records, for instance, are based on self-report, with no guarantee of validity. Diagnoses, which may serve as both independent (predictor) and dependent (criterion) variables, may be biased and even influenced by expectation or pre-existing beliefs about the relationships between diagnoses. When that is the case, any conclusions drawn from a study's findings will not generalize to patients with verifiable diagnoses.

2.1.4 Selection Bias and Random Assignment

Another important threat to internal validity is bias in selection or assignment to groups. Random assignment should eliminate systematic, pre-existing differences between experimental groups, which is why it is a component of a true experiment. When participants are not randomly assigned to groups (e.g. intervention versus control group), there is a risk that the groups will be different to begin with, independent of the variable of interest. To illustrate, if one is comparing the effect of exercise on cognition, and participants are allowed to self-select, the more energetic individuals may choose the exercise group while the less energetic individuals choose the alternative option of, for example, computer games. When measuring cognition upon completion of the study period, the exercise group may perform better than the computer game group, but this may be due to pre-existing differences in energy and not due to the addition of exercise. Perhaps, those with low energy are afflicted with a condition, a third variable that impacts cognition. Conceivably, this could be a condition such as diabetes or sleep apnea, each of which can impact both energy and cognition.

Group differences may also be impacted by pre-existing differences if individuals are assigned to levels of the independent variable or intervention according to a nonrandom factor. For instance, say the researcher is interested in evaluating a new reading intervention for third graders. The researcher has the teachers in two classes in the school provide the standard reading curriculum while the teachers in two additional classes administer the new reading intervention. Upon conclusion of the school year, the researcher compares reading scores between the two sets of classes and finds that the intervention classes have higher reading scores. Unfortunately, they cannot conclude that the intervention caused the higher test scores because the two sets of classes may have differed in some way(s) other than the reading curriculum. The two sets of classes may have differed on some characteristic like attention or parental involvement. It is not uncommon for school administrators and counselors to assign certain types of children to specific teachers depending on their ability to work with those children. Some parents may go out of their way to request a particular teacher for their child prior to the start of the school year based on what they know about the teacher. Perhaps, the teachers in the two sets of classes have different levels of experience. One can think of a lot of ways the classes may differ independent of the reading intervention. A better strategy would be to randomly assign participants to teaching method independent of their class placement or teacher. Alternatively, a researcher could expand their study across more schools and more teachers, while randomly assigning the reading curricula to the various classes. It is possible to manage so-called nuisance variables (e.g. school, teacher, or any variable not directly of interest) with specific research designs, such as Latin square designs and nesting. These will be reviewed more in Chapter 3.

Sometimes, even with random assignment, groups may differ slightly with respect to an extraneous variable (e.g. age, education). This can be managed by treating that variable as a second factor to be analyzed for its significance. Alternatively, the effects of the extraneous variable can be partialled out statistically by treating the variable as a covariate with analysis of covariance or partial correlation. These design strategies are reviewed in more detail in Chapters 4 and 5.

2.1.5 Attrition

Experimental mortality, or attrition, can also threaten the internal validity of a study. Actual mortality may occur in certain types of medical research, but, in general, we are talking about participants dropping out of a study for various reasons. If we are comparing two groups on an outcome, and there is more attrition in one group than the other, this could affect the results of our data analysis. The analysis could indicate differences between the groups that are not related directly to the variable of interest. In some cases, attrition may favor your research predictions, while in other cases the attrition could wash out real effects that may have occurred.

Let's say we want to compare an antidepressant medication to cognitive behavioral therapy (CBT) in the treatment of anxiety. We start with 50 patients in the medication group, and 50 in the CBT group. By the end of the six-month study period, half of the medication patients have dropped out of the study, and only two patients have dropped out of the CBT arm of the study. We run our analyses, and the 25 patients in the medication group are found to have statistically lower scores on a measure of anxiety compared to the CBT group. We conclude that medication was superior to CBT. In fact, we could be wrong. An alternative explanation is that the people who dropped out of the medication group did so due to worsening of their symptoms and we were left with a select group of patients who would have experienced improvement due to the passage of time or some other factors unrelated to the medication. The medication may not have helped at all, or the medication may have only been beneficial for patients with certain characteristics. Further investigation would be needed to determine why the participants dropped out.

2.1.6 Experimenter Bias

Closely related to the issue of nonrandom assignment is experimenter bias. Experimenter bias is the phenomenon by which the researcher influences the outcome of the study by their own biases. This is usually an unconscious bias, but a significant threat to internal validity nonetheless. Experimenter bias may occur in the administration of the intervention, or in the assessment of the outcome. The researcher's administration of the intervention may be influenced by their enthusiasm or their experience in using the intervention. Their assessment of outcome may be influenced by their belief about the effectiveness of the intervention(s).

Experimenter bias may be managed on the assessment end by using blinding. In this case, the person measuring the outcome is blind to, or unaware of, group assignment. In other words, the person measuring the dependent variable should not know which intervention was received by the participant that they are assessing on the outcome measure.

On the intervention end, blinding may be possible in certain types of studies, such as studies comparing one medication to another, or one medication to a placebo pill. In that case, the researcher or clinician is not aware of which medication (or placebo) is being taken by each participant. In other types of intervention studies, it is practically impossible to blind the research personnel. For instance, if you are comparing CBT versus psychoanalytic therapy, the therapist is obviously going to know which type of therapy they are administering. In this type of situation, it is best to have multiple therapists across multiple sites, and at least try to blind the therapists to the predictions of the lead investigator.

2.1.7 Expectation

Expectation on the part of the participant can additionally impact internal validity. The participants' belief that one type of intervention will be effective may influence their response to that intervention. This is referred to as a placebo effect when they expect a medication or treatment to improve their symptoms. This can be addressed by including a placebo-control group and blinding both the intervention group and the control group as to whether they are receiving a real treatment versus a placebo (e.g. sugar pill or non-medicated patch). In other types of studies, the participant can be blinded to the purpose of the study; meaning, they are not told whether they are in the treatment arm that is expected to be more beneficial versus the one that is not expected to be beneficial.

2.1.8 Sensitization and Practice Effects

Finally, internal validity can be impacted by pretesting. That is, assessment of baseline functioning before an intervention may influence performance on the assessment measure following the intervention. This may be due to the effects of repeated measurements, or the sensitization caused by the initial performance.

In terms of repeated testing, practice effects may influence later test scores. This may be managed by using alternate test forms, though that is not ideal since alternate measures rarely have perfect test–retest relatability and, therefore, measurement error will be introduced into the data analyses. A better way to manage the impact of repeated testing is to include a control group that also completes both baseline testing and testing at the conclusion of the study period. A larger or smaller change over the study period in the intervention group compared to the control group suggests that the change occurred due to a factor other than practice effects. That is, if the change from pretest to posttest is greater for the intervention group than the change in scores for the control group, a researcher can conclude with some degree of confidence that the change was related to group membership (e.g. treatment) and not just due to practice effects.

Sensitization caused by pretest, or baseline, assessment may be more difficult to manage. One strategy is to make sure the participant is blind to the predictions of the study. Additionally, the participant should not know how they performed on the baseline test or what the initial assessment indicated. This is one of those situations in which face validity would not be a good thing.

2.1.9 Incorrect Conclusions of Causality

As noted at the beginning of this chapter, threats to internal validity impact the researcher's ability to confidently conclude that there is a causal relationship between the events or conditions under study. If there are flaws in the design, or if the research is not truly experimental in nature, causality cannot be concluded. Recall that a true experiment is one in which there is random assignment to groups, manipulation of the experimental variable (i.e. the experimental variable is created, not innate), and the researcher has controlled the extraneous influences that may affect the outcome.

In Table 2.1, I give examples of studies that are not truly experimental in nature. In each case, I provide an incorrect statement about causality followed by the correct conclusion pertaining to the study results.

2.2 External Validity

External validity also affects a researcher's ability to draw conclusions about a study's results. External validity is the degree to which a study's results generalize. In other words, do the results apply to situations and populations outside of the research? Two aspects of generalization relate to *Who* and *Where*.

First is the Who: do the study results apply to individuals and groups other than those in the study? The answer to this question will depend on how well the participants represent the general population. The broader and more diverse the study sample, the better the results will apply to the general population. But if the study sample is restricted to certain demographics, the results will only generalize to individuals and groups sharing those demographic characteristics. Important characteristics include, but are not limited to, gender, ethnicity/race, country of origin, language, age, hand preference, education, and economic factors. Other characteristics will have even more importance depending on the nature and purpose of the study.

Characteristics influencing generalizability are not limited to demographics. For instance, in treatment studies comparing one treatment method to another, disease severity could influence generalizability. In other words, the effectiveness of one treatment method over another may vary according to baseline

Table 2.1 Erroneous statements of causality.

Study design[a]	Incorrect statement of causality[a]	Correct description and alternative explanations[a]
Medical records were reviewed for history of cancer and Vitamin D supplementation	Vitamin D supplementation causes cancer. Error: This was not an experimental design. The independent variable, Vitamin D supplementation, was not manipulated; participants were not randomly assigned to Vitamin D supplementation versus no supplementation to see if this results in differences in cancer rate.	There was an association found between cancer diagnosis and Vitamin D prescriptions. It is unclear if Vitamin D caused cancer, or if physicians were more likely to prescribe Vitamin D to patients at risk for cancer
People in a phone survey who reported consuming two or more glasses of red wine per day had less heart disease than people consuming no red wine	Red wine protects against heart disease. Error: This was not an experimental design. Participants were not randomly assigned to wine consumption versus no consumption groups.	There was an association between reported red wine consumption and less heart disease. It is unclear if this was causal, or if this was due to another factor, such as healthy people were more likely to consume wine, or people who were socially engaged were more likely to drink wine, with social engagement improving heart health. Additionally, if there was a true causal relationship, it may have been a component of the red wine (e.g. the anthocyanins in the grapes), and not the alcohol, that reduced heart disease
Standardized math scores and data on participation in elective music classes were obtained from school records	Strong math ability leads to more participation in music activities. Error: This was not an experimental design. Math ability is a quasi-experimental variable; it cannot be manipulated.	There was a relationship between math scores and participation in music classes. It is unclear whether this was due to a relationship between general cognitive skills and music, a relationship between lack of need to study resulting in more time to participate in extra classes, or other third variables. Similarly, if there is a causal link between math ability and music activities, the direction of the causality could go either way; that is, musicality could improve math skills *or* vice versa

Self-report data were obtained for time spent watching TV along with teachers' reports of incidents of aggression among high-school students	Watching TV causes aggression. Error: This was not an experimental design. TV watching was not manipulated; participants were not randomly assigned to high, medium, or low TV viewing time (for example).	Increased time watching TV was associated with more incidents of aggression. It is unclear whether this was causal, or if aggressive adolescents were more likely to watch TV versus engaging in other activities. Aggressive individuals may have been less likely to be invited to participate in group activities and, thus, had more time to watch TV. Families that promoted TV watching may have also been less likely to promote prosocial behavior. Families with aggressive children may manage their children's behavior by keeping them entertained with TV
Students reported the amount of time they spent studying for the SAT and their final SAT scores	SAT preparation increases SAT scores. Error: This was not an experimental design. Students of all levels of aptitude were not randomly assigned to groups with time spent studying as an independent (manipulated/created) variable (e.g. a group assigned to study 0–1 hours per week; a group assigned to study 1–5 hours per week; and a third group assigned to study 5–10 hours per week).	There was a positive relationship between time spent preparing for the SAT and SAT scores. It is unclear if this was due to a causal relationship, or if students who perform well on standardized aptitude tests were also more likely to study; students motivated to do well on the SAT in order to attend more competitive universities may have had higher levels of innate aptitude unrelated to study time
Records of primary care patients were reviewed for rates of participation in group therapy and suicide attempts	Participating in group therapy led to an increased likelihood of suicide. Error: This was not an experimental design. Participants were not randomly assigned to group therapy versus other treatment for comparison.	There was a positive association between participation in group therapy and suicide attempts. It was not possible to determine the direction or presence of causality. It is possible that depressed patients are more likely to be referred to group therapy and depressed patients are more likely to attempt suicide

(Continued)

Table 2.1 (Continued)

Study design[a]	Incorrect statement of causality[a]	Correct description and alternative explanations[a]
Patients going through cancer treatment were found to have lower scores on measures of working memory	Chemotherapy causes diminished working memory. Error: This was not an experimental design. Participants could not ethically be randomly assigned to chemotherapy versus other treatment. Extraneous variables were not measured or controlled; these included sleep disruption, cancer type, activity level, mood and anxiety, and distractions. Participants may not have been blinded to the experimental hypotheses, and they may have had expectations about decreased working memory that may have influenced their test performances.	Patients in a cancer clinic had lower scores on a measure of working memory than their non-patient peers. It is not clear whether this was due to chemotherapeutic agents, sleep disruption due to multiple factors, the cancer itself, inactivity, depression or anxiety, distraction, expectation bias, or other factors that were not controlled
Patients participated in three weeks of education about the benefits of exercise on mood	Exercise education improved mood. Error: Without a control group, the effects of history and maturation could not be managed; improvement may have resulted from other events or natural recovery.	Patients' mood was better after three weeks of exercise education, but causality could not be determined since there was no control group and since actual exercise was not measured. Mood may have improved due to historical events, such as decreased cloudiness or other events. Mood may have also improved due to the attention and social interaction received during the study. Finally, mood may have improved spontaneously as a natural course of their condition
Size of attendance at sporting events was recorded along with win–loss records	Higher attendance rates contributed to a better win–loss record. Error: This was not an experimental design. Attendance was not manipulated with attendance levels randomly assigned to teams.	There was an association between attendance rates and win–loss records. It is not clear whether high attendance resulted in better performance by teams, or if better teams attracted more attendees. Causality could only be determined if researchers manipulated attendance and then measured win–loss records prospectively

Study	Conclusion / Error	Explanation
Figure skaters participated in yoga during the season, and the number of falls during a routine at the end of the season was compared to the number of falls at the beginning of the season	Participation in yoga improved jumping success. Error: Maturation was not managed by having a comparison control group that did not participate in yoga. Natural maturation and other factors may have contributed to a decrease in falling.	Jumping success improved among skaters. Yoga may have contributed to that success, but it is impossible to conclude causality without a comparison group that did not participate in yoga. Other factors, including maturation, training, a change in technique, or a change in coaching may have improved success
One hundred brains of professional athletes in contact sports were submitted by family members for autopsy to test for chronic traumatic encephalopathy (CTE), and 99 were determined to have the pathology of CTE	Participation in contact sports causes CTE almost universally. Error: Participants (deceased athletes) were not randomly selected from all deceased athletes. We do not know, therefore, what percentage of athletes have CTE pathology due to the presence of a select, or potentially biased, sample. Similarly, there is no comparison control group of non-athletes for comparison purposes to control for natural maturation.	A causal link between contact sports and CTE could not be concluded based on these data alone due to issues of selection bias. The findings could not be generalized to all contact sport participants since the sample was not randomly selected. There were also no baseline data for comparison to determine if CTE was present before participation in contact sports, or only following participation. The relationship between the number of head injuries and degree of pathology would also be helpful in interpreting the data
Performance on state-mandated measures of achievement were higher at School "A" than at School "B"	Teaching performance at School "A" is better than at School "B." Error: This is not an experimental design. Students were not randomly assigned to schools.	Students at school "A" outperformed those at School "B." There were many factors that may have contributed. While teaching methods may have been a factor, this was impossible to establish since the students may have differed at baseline, experiences outside of the school setting may have influenced achievement, and other factors within the school besides teaching may have influenced performance

a) Most of these are completely made-up studies.

disease severity such that the one treatment may be effective for patients with one level of severity (e.g. mild), but not for those with more severe or advanced disease states.

As important as being able to generalize to the whole population is the ability to generalize outside of the laboratory setting. For instance, participants may react differently because they are in a laboratory setting (reactive effects). They may also respond differently just knowing they are part of a research study, particularly if they know the purpose of the study (expectation). This also poses a threat to the internal validity of the study. As noted earlier in this chapter, this issue can be addressed to some extent by blinding the participants to the purpose of the study.

Since participants may react to the experimenter's bias or expectations (i.e. experimenter effects), a second strategy as discussed earlier is to blind the experimenter to the participant's condition. This is most applicable to drug studies in which neither the participant nor the person measuring response to treatment are made aware of which drug has been administered. Alternatively, for nondrug studies, interventions or experimental manipulations may be administered by non-investigators or neutral parties.

An additional strategy for improving generalizability is to conduct studies in naturalistic settings. For instance, instead of bringing participants into a laboratory setting, they may be observed in the real world. In some cases, it may even be possible to implement an intervention or manipulation of the environment to evaluate people's reaction to the change. As discussed in Chapter 1, though, it is important to ensure that any observations made or manipulations applied to the environment or situation are ethical and pose no harm. Before these studies are conducted, it is critical to have the methods reviewed and approved by the appropriate local authority (e.g. the Institutional Review Board).

Being influenced by the pretesting was noted earlier to have potential consequences for internal validity. This phenomenon can also be thought of as affecting external validity. Again, this issue can be addressed by including at least two groups (e.g. experimental and control groups).

Closely related to issues with pretest sensitization are practice effects and carryover effects. Practice effects refer to the impact of pretesting on post-testing attributable to experience with the pretest. That is, simply taking the test a second time may result in improved performance due to familiarity with the test. Practice effects can be managed by using alternate test forms and/or having a control group as discussed in Section 2.1.8.

Carryover effects refer to a phenomenon that occurs in studies in which participants experience more than one research condition. Those studies are characterized as being within-subjects designs, to be discussed in Chapter 3. In these studies, participants' response to one intervention, for example, is compared to their response to a second intervention. The problem is that there may be carryover in their response from one condition to the other. That is, the reaction to the first condition may still be present when they are exposed to the second condition. Carryover effects may also be thought of as order effects.

Carryover effects can be managed by alternating the order in which participants are exposed to one condition versus the other. That is, one group of participants can receive Intervention/Condition "A" first, followed by Intervention/Condition "B," while a second group is exposed to Condition "B" first and then Condition "A." As the number of conditions being considered increases, so too will the number of possible orders.

Carryover effects are not an issue if each participant is exposed to only one condition. This would be the case in a between-subjects study design (discussed in Chapter 3) in which one group of participants is exposed to one condition or treatment, and another group is exposed to the other condition or treatment. The advantage of the within-subjects design in which each participant is exposed to each condition, though, is that participants serve as their own controls, managing the effects of extraneous variables (age, gender, pre-existing tendencies). It is possible to have the best of both worlds by having a within-subjects variable (i.e. by having each participant exposed to each treatment) and by controlling for order as described in the previous paragraph, and including order as a between-subjects variable. The design has now become a between–within subjects design, with condition serving as a within subjects factor, and order as a between subjects factor. This will be clarified in Chapter 3.

In Table 2.2, I present examples of studies with poor external validity. Inclusive, I review the errors and the correct interpretation.

Table 2.2 Erroneous statements of generalizability.

Study design[a]	Incorrect statement of causality[a]	Correct description and alternative explanations[a]
Hospitalized patients at a psychiatric facility were treated with music therapy, and their level of depression was assessed before and after two weeks of music therapy.	Music therapy improves depression. Error: The sample studied was limited to patients with depression severe enough for hospitalization, limiting generalizability. There also may have been natural improvement (maturation) or improvement due to other (extraneous) factors. Experimenter bias and expectation may have also influenced the findings. Use of a control group may have resulted in improved ability to deduce causality between music therapy and improvement, though generalizability would still be limited to patients with severe depression.	Patients with severe depression showed improvement in depression across a two-week period during which they received inpatient treatment that included music therapy. Depression may have responded to overall treatment and or the passage of time.

(Continued)

Table 2.2 (Continued)

Study design[a]	Incorrect statement of causality[a]	Correct description and alternative explanations[a]
A researcher conducted a study to see if instructing parents on the benefits of reading with their children improves reading scores in children. The researcher compared two groups at a school in an upper, middle-class neighborhood; the parents of the children in one group received the special instruction, and the parents of the comparison group did not. Children were randomly assigned to one group or the other.	Providing instruction to parents on the benefit of reading with their children will improve reading performance in those children. Error: While this study represents an experimental design with random assignment and manipulation of the independent variable (instruction in the benefits of reading with children), the findings may not generalize to all children and parents given the possible restriction in parent/child demographics.	Among children in an upper middle-class neighborhood, parental instruction regarding the benefits of reading with their children resulted in improved reading scores among the children. This study should be replicated in schools with different demographics to determine whether the instruction is beneficial across the general population.
Participants kept a headache diary to assess the impact of caffeine withdrawal. They consumed their usual caffeine for a week, followed by a week of no caffeine.	Average weekly headache rating increased due to caffeine withdrawal. Error: Participants were not blinded to their condition (caffeine or no caffeine), and so their expectation of caffeine withdrawal may have biased their headache ratings, which were subjective. This may have been exacerbated by the fact that they were enrolled in a study – therefore, the results may not have generalized to their everyday, naturalistic environment.	Participants who were instructed to monitor their headaches with and without caffeine, self-reported more headaches when they knew they were not consuming caffeine. This may have been due to either physiological or psychological factors, and the findings may also have been influenced by their focus on headaches due to being enrolled in a study on caffeine and headaches.

Table 2.2 (Continued)

Study design[a]	Incorrect statement of causality[a]	Correct description and alternative explanations[a]
College students were recruited to participate in a study of the effect of violent video games on aggression. Half the students were randomly assigned to playing a first-person shooter game, and half were assigned to playing an animated construction game. An inflatable doll was placed in the room with each participant, and a researcher recorded the number of times the player struck the doll.	Playing violent video games increases aggression. Error: While this study represents an experimental design for the most part, with random assignment and manipulation of the independent variable (type of video game), the results of the study may not generalize outside of the laboratory setting. Inclusive, the "aggression" may not generalize to physical aggression on humans or other living beings or to objects that can be damaged. Additionally, because the participants were aware this was a study of the effect of video games on aggression, they may have reacted to their own expectation of increased aggression while playing a violent video game. Related, the researchers who recorded acts of aggression may have been biased in their recordings by their beliefs about violent video games and aggression. Finally, the demographics of the sample were somewhat restricted, and the findings may not generalize to the broader population or across age groups.	Within a laboratory setting, college students playing a first-person shooter showed more aggression toward an inanimate, indestructible toy than students engaged in playing a construction video game. It is not clear if the aggression would apply to real-world situations where harm or damage could be inflicted, and it is not clear whether the students were responding to expectations of the researcher. Finally, it is not clear whether the same findings would hold among younger or older groups, or those with less education.

a) Most of these are completely made-up studies.

2.3 Summary

The ability to draw conclusions from the results of a research study will depend on the study's validity. Validity can be characterized somewhat as falling into one of two broad categories, internal and external validity. Internal validity impacts the degree to which a researcher or consumer of research can conclude that there is a causal link between the factors under study. External validity basically determines the degree to which a research study's findings extend outside of the study setting and to what groups or segments of the population. Many aspects of internal and external validity can be managed through careful design of the research methods.

In Chapter 3, I review some basic research designs. Many of these study designs lend themselves well to statistical techniques that evaluate group differences or changes over time. Those statistical techniques along with techniques for predicting group membership or outcome are discussed in Section 3.

3

Research Designs

In this chapter, I review research designs that can be used to test hypotheses. Research designs can be categorized as either experimental or nonexperimental, with quasi-experimental designs falling between the two. As discussed earlier, a true experiment is one in which the independent variable is manipulated, participants are randomly assigned to one condition or another, and extraneous factors are controlled as much as possible. A quasi-experimental design meets these criteria with the exception that the independent variable, or condition of interest, cannot be manipulated because it is innate or fixed. Examples of quasi-experimental independent variables are gender, hand preference, ethnicity, and country of origin. In nonexperimental designs, participants may not be randomly assigned to conditions. Instead, participants select the condition in which they participate, or convenience samples are used in which participants already belong to one condition or another prior to the inception of the study. Nonexperimental designs are common in epidemiological studies.

3.1   The Lingo

When we talk about research designs, we tend to use a variety of terms. Whatever design and statistical analysis we use, two important concepts are *independent variable* and *dependent variable*. Some researchers will alternatively use the terms *predictor* and *criterion* variable, particularly if the research design does not permit conclusions about causality. In this primer, I will often use the former terms, regardless of causality.

Independent variables are the conditions the researcher is interested in evaluating, while the dependent variables are, loosely speaking, the outcome measures. For instance, if I am interested in comparing two different interventions,

Research Design and Analysis: A Primer for the Non-Statistician, First Edition. Leslie D. Rosenstein.
© 2019 John Wiley & Sons, Inc. Published 2019 by John Wiley & Sons, Inc.

intervention is my independent variable. The dependent variable is the characteristic that I hope my intervention will impact. In the case of a study comparing two interventions in the treatment of depression, for example, my dependent variable would be some measure of depression.

In the above example, if I am comparing three different interventions, then my independent variable has three levels. If I am comparing a drug to a placebo, then my independent variable, treatment, has two levels. The levels are (1) drug "X" and (2) placebo. I can also have more than one independent variable. For instance, I may be interested in how men and women differ in response to the new drug versus the placebo. Now, I say I have two factors of interest, gender and treatment. Each factor has two levels. A study can have any number of independent variables, or factors, each with any number of levels. However, the more independent variables and the more levels of the independent variables I have, the more participants I will need in order to analyze my data.

When a researcher is interested in one factor alone, they have a single-factor design. When more than one factor is considered in a study, the design is referred to as a multifactorial design. Most of the following designs refer to studies comparing average scores, or means, across groups on one or more dependent variable(s).

A final piece of terminology that has to be discussed before moving on to a discussion of specific research designs is the concept of error variance. Error variance is any variability in the dependent variable (scores) that is due to anything other than our factors, or independent variables. One reason error variance is important to understand is that a significant finding in tests like analysis of variance (ANOVA, to be covered in Chapter 5) indicates that the variation across groups (i.e. the independent variable(s)) is significantly more than the variation, or variance, within the groups (the subjects). So, any variance across *subjects* (within the groups) that is not related to the independent variable will reduce our ability to detect an effect of that independent variable. To put it another way, we are interested in differences, or variation, from one group to another, which is between group variance. For differences across groups to be significant, those differences must be significantly larger than the differences across subjects, within each group.

Now that you have some of the lingo down, it is possible to talk about the good stuff – the actual designs. As noted in Chapter 2, most of the designs I review here lend themselves to statistical analyses that test differences between groups or change over time (within groups). These designs can be classified into between-subjects designs, within-subjects deigns, and between–within subjects designs. I will also review methods for implementing these designs and assigning participants to groups to maximize efficiency without sacrificing design integrity.

3.2   Between-Subjects Designs

A between-subjects design is one in which the researcher compares two or more groups of participants on a dependent variable. In a single-factor design, the researcher may compare participants who have been divided into two or more groups on the basis of one factor, or independent variable. For instance, the researcher may look at math aptitude (the dependent variable) across groups of left-handers (LH), right-handers (RH), and ambidextrous individuals (AH). The independent variable is hand preference, with three levels (LH, RH, and AH).

Between-subjects designs can also come in the form of a multifactorial study with two or more independent variables. Let's take the previous example of hand preference and math aptitude. Let's say that we think left-handers will have higher math aptitude than right-handers and ambidextrous individuals, but we think that may also depend on gender. So, now we divide our hand preference groups further into male and female groups. We now have two independent variables, hand preference and gender. One independent variable, or factor, has three levels (hand preference), while the other has two levels (gender). We still have one dependent variable (score on a math aptitude test). This study can be summarized as a 3×2 between-subjects factorial design. Incidentally, too, this study is quasi-experimental, since the independent variables cannot be manipulated; that is, participants' hand preference and gender have been determined prior to the inception of the study.

3.2.1   More Examples of Between-Subjects Designs

In Table 3.1, I describe a single-factor between-subjects design with two levels of the independent variable.

Table 3.1 Single-factor between-subjects design with two levels of the independent variable.

Independent variable: Treatment
    Levels:
        1: Stimulant drug
        2: Standard treatment

Dependent variable: Number of accident injuries over the course of two years
50 adult patients with ADHD are treated with a stimulant medication. 50 adult patients with ADHD are treated with standard care without the addition of a stimulant medication.
After a two-year period, the number of accident injuries reported to the primary-care physician (PCP) during appointments is totaled.

Table 3.2 Single-factor between-subjects design with four levels of the independent variable.

Independent variable: Treatment
Levels:
1: Treatment with a stimulant drug
2: Treatment with a selective norepinephrine reuptake inhibitor
3: Computerized inhibition training
4: Standard treatment (i.e. treatment as usual)
Dependent variable: Number of accident injuries over the course of two years
50 adult patients with ADHD are treated with a stimulant medication.
50 adult patients with ADHD are treated with a selective norepinephrine reuptake inhibitor.
50 adult patients with ADHD participate in a computerized training program designed to improve response inhibition.
50 adult patients with ADHD are treated as usual by their PCP.
After a two-year period, the number of accident injuries reported to the PCP during appointments is totaled.

In Table 3.2, I describe a single-factor between-subjects design with four levels of the independent variable. This study also has only one dependent variable.

In Table 3.3, I describe a single-factor between-subjects design with three levels of the independent variable. In this study, there are three dependent variables. This study can be analyzed with a multivariate analysis of variance (MANOVA; to be discussed in Chapter 5).

In Table 3.4, I illustrate a multifactorial, between-subjects design with two independent variables and one dependent variable. One of the independent variables (school) has two levels, while the other has three levels (major). Because we have more than one independent variable, we will be able to evaluate the interaction between them. The concept of interaction will be explored further in Chapter 5. Briefly, though, an interaction is an interplay between two independent variables such that the impact, or effect, of one independent variable is modified by the impact of another independent variable. Put another way, the impact of one independent variable will differ depending on the level of the second independent variable.

3.2.2 Statistical Analyses for Between-Subjects Designs

As noted earlier, between-subjects designs result in data that can be analyzed with tests assessing differences between groups. When you have a single-factor design, with just two levels of the one independent variable and your

Table 3.3 Single-factor between-subjects design with three levels of the independent variable and three dependent variables.

Independent variable: Treatment
    Levels:
        1: Treatment with a stimulant drug
        2: Treatment with a selective norepinephrine reuptake inhibitor
        3: Standard treatment (i.e. treatment as usual)

Dependent variables:
    1: Number of accident injuries over the course of two years
    2: Number of Emergency Department (ED) visits
    3: Number of pain medication prescriptions or refills

50 adult patients with ADHD are treated with a stimulant medication.
50 adult patients with ADHD are treated with a selective norepinephrine reuptake inhibitor.
50 adult patients with ADHD are treated as usual by their primary care physician (PCP).

After a two-year period, the number of accident injuries reported to the PCP during appointments is totaled along with the number of ED visits for each participant, and the number of pain medication prescriptions of refills.

Table 3.4 A 2×3 multifactorial between-subjects design with two independent variables.

Independent variables: Treatment and gender
    Treatment with two levels:
        1: Treatment with a stimulant drug
        2: Standard treatment (i.e. treatment as usual)
    Gender with two levels:
        1: Male
        2: Female

Dependent variable: Number of accident injuries over the course of two years

25 adult male patients with ADHD are treated with a stimulant medication.
25 adult female patients with ADHD are treated with a stimulant medication.
25 adult male patients with ADHD are treated as usual by their primary care physician.
25 adult female patients with ADHD are treated as usual by their primary care physician.

After a two-year period, the number of accident injuries reported to the primary care physician during appointments is totaled.

---

dependent variable is a continuous variable (like height, age, and test score), you can conduct a *t* test or an ANOVA or one of their nonparametric equivalents (to be reviewed in Chapter 6). So, if you have one independent variable, say hand preference, and that independent variable has just two levels in your study, say right-handedness and non-right-handedness, and your dependent variable is measurable on a continuum, then you can opt for either a *t* test *or* an ANOVA (or related nonparametric test) to test for group differences on the dependent variable, such as handwriting quality.

But, now let's say you want to divide your non-right-handers into groups of left-handers and ambidextrous participants, so you now have three levels of handedness (right-handers, ambidextrous participants, and left-handers). Because you have more than two levels of your independent variable, you will use ANOVA and not a *t* test to analyze your data for group differences. That is because *t* tests are for use with two groups, while ANOVA is used when you have two *or more* groups.

Note that ANOVA has many variations, including analysis of covariance (ANCOVA), which allows the researcher to statistically control for the effects of a measured extraneous variable. Another variation of ANOVA involves the number of dependent variables that can be considered in a single analysis; when the researcher assesses the effect of the independent variable(s) on more than one, related dependent variables, the data can be analyzed with MANOVA.

t tests, ANOVA, ANCOVA, and MANOVA are reviewed in more detail in Chapter 5. The nonparametric equivalents of *t* tests and ANOVA are reviewed in Chapter 6.

3.3   Within-Subjects Designs/Repeated Measures

In contrast to a pure between-subjects design, a pure within-subjects design is one in which all participants are assigned to one group, but the dependent variable is measured at multiple points. In an intervention study, for example, the participants may undergo pretesting to obtain their baseline, and then repeat testing following the intervention. This would be a single-factor, within-subjects design. Or, if there are two (or more) levels of the intervention, each and every participant may be exposed to two (or more) interventions. For instance, a researcher may be interested in comparing cognitive behavioral therapy (CBT) to psychodynamic therapy (PsyD) in the treatment of anxiety. Every participant may be treated for six months with PsyD and six months of CBT. This would also be a single-factor within-subjects design. In this case, the dependent variable may be measured twice – upon completion of each intervention, or thrice – at baseline and again following each intervention. This study can also be described as a crossover design – each participant crosses over from one condition to the other condition.

But wait, you say. What about carryover effects? If participants tend to improve after PsyD, how will the researcher know whether CBT had any benefit? The effects of the PsyD may still be present when they participate in CBT.

Table 3.5 A within-subjects design with three levels of one within-subjects independent variable.

Independent variable: Time

    Time with three levels:

        1: Baseline

        2: Immediately post-treatment

        3: Follow-up

Dependent variable: Score on a depression questionnaire

50 patients referred to a depression clinic are treated with cognitive behavioral therapy (CBT).

The patients complete a depression questionnaire prior to starting treatment, again immediately following the designated six-month treatment period, and again at follow-up six months later.

As discussed earlier in Chapter 2, it is best to control for carryover effects by assigning participants to one of two (or more) orders. Half the participants in the previous example would receive PsyD therapy for six months followed by six months of CBT, while the other half would receive CBT first, followed by PsyD. And, to be really thorough, the researcher can even analyze the effects of order by treating order as an independent variable. Order is a between-subjects variable, because one *group* is assigned to the first order, and a separate *group* is assigned to the second order. This is an example of a between–within subjects design. In this case, the researcher has a 2 × 2 between–within subjects factorial design, with two levels of order (the between-subjects variable) and two levels of therapy type (the within-subjects variable).

In Table 3.5, I outline a within-subjects design with one independent variable with three levels. This is a study of the effectiveness of CBT in the treatment of depression. Even though my variable of interest is the CBT, my independent variable is Time. That is because in this within-subjects design, I am interested in the change in depression over time, from baseline to post-treatment and later follow-up. The dependent variable is the level of depression as measured by a depression questionnaire. Ultimately, I am attempting to measure response to treatment (though we know from our discussion of threats to internal validity, that without a control group, I could also be measuring maturation and/or response to history). In a later example, I will demonstrate how to design a study to manage those threats.

3.3.1 Statistical Analyses for Within-Subjects Designs

As with between-subjects designs, a study conducted with a within-subjects design results in data that can be analyzed with tests assessing differences, but this time the difference is across treatments or time within the same individuals. Again, when you have a single-factor design, with just two levels of the one independent variable (e.g. placebo versus drug), you can conduct a *t* test. The *t* test for a repeated measure, or within-subjects design, is slightly different than the *t* test for a between-subjects design. Instead of an independent *t* test, you will need to use a matched pairs *t* test (also referred to as a dependent *t* test). When you have three or more levels of the within-subjects variable, you can use an ANOVA assessing change over time (e.g. ANOVA-repeated measures) or Mixed Linear Model analysis for repeated measures. The matched pairs *t* test, ANOVA-repeated measures, and Mixed Linear Model analyses are reviewed in more detail in Chapter 5.

3.4 Between–Within Subjects Designs (Mixed Factorial/Split-Plot Designs)

A between–within subjects design includes both between-subjects variables (e.g. hand preference, gender) and repeated measures (e.g. Apparatus 1 vs. Apparatus 2; Pre-, Post-, and Follow-up Treatment performance; or Treatment A versus Treatment B versus Treatment C). These studies are also sometimes referred to as mixed factorial designs or split-plot designs.

As mentioned earlier, a strict within-subjects design is not robust against the threats to internal validity posed by issues such as maturation and history. Those threats, though, can be managed by adding a control group. So, instead of assigning all participants to one group, such as music therapy, half the participants can be assigned to the music therapy group, and half can be assigned to treatment as usual. Music therapy treatment versus treatment as usual now represents a between-subjects variable with two levels, and both groups can be tested over time (before treatment, immediately following treatment, and again at follow-up). This is a between–within subjects design, and the between-subjects variable allows you to manage the effects of maturation and history as both groups should experience roughly the same experience with the same passage of time.

A drug trial may be designed as a between–within subjects design with drug versus placebo serving as the between-subjects variable, and time serving as the within-subjects variable (e.g. baseline, post-treatment, and follow-up). In this case, half the participants would be randomly assigned to the drug treatment, and the other half would be assigned to the placebo control group. The dependent variable (i.e. drug response) would be measured at the three time points for all participants in both groups.

When we conduct this type of between–within subjects design, we are typically most interested in the interaction between the between-subjects variable and the within-subjects variable. That is, we want to know if there is a relative difference between our two groups (treatment versus control, the between-subjects factor) in their change across time (the within-subjects factor). For instance, I have a treatment group and a waitlist control group, and I test their hemoglobin A1C before the intervention, after the intervention, and again at a follow-up appointment. In this study, I am hoping that the treatment group will have more improvement over time relative to the placebo control group; that would confirm my hypothesis that the treatment improved hemoglobin A1C above and beyond what might have occurred just due to the passage of time or standard care.

In Table 3.6, I detail a study using a between–within subjects design with three between-subjects variables and one within-subjects variable. There is one dependent variable. The two between-subjects variables of interest are gender with two levels, and age group with three levels. The within-subjects variable is treatment modality, with two levels – medication and CBT. I have also included Order as a (third) between-subjects variable so that I can directly evaluate whether the relative effect of the two treatment modalities is impacted by the order in which each treatment is received. The dependent variable is score on a depression questionnaire. In this study, I am interested in the relative effects of medication versus CBT on depression in men and women at different age levels.

3.4.1 Statistical Analyses for Between–Within Subjects Designs

When the researcher is primarily interested in differences between and within groups, the data from a study with a between–within subjects design can be analyzed using repeated measures analysis of variance (ANOVA-RM) or Mixed Linear Model analysis. These procedures and when or whether to use each are reviewed in Chapter 5.

In the following sections, I begin reviewing ways to maintain design integrity and study validity while maximizing efficiency. The designs in the following sections are variations on between-, within-, and between–within subjects designs. In each case, subjects, or participants, are assigned to experimental groups in a manner that minimizes the number of participants required without creating confounds between the variables of interest and extraneous variables. First, I will introduce Latin square designs. You might want to skip this section unless this is an area of particular interest to you because it is somewhat esoteric. Latin square designs are a special type of design that allow the investigator to assign participants to treatment groups while minimizing the number of participants required for the study.

Table 3.6 A 2×3×2×2 between–within subjects design with one dependent variable.

Independent variables: Gender, Age group, and Treatment modality
  Three independent between-subjects variables:
    Gender:
      1: Male
      2: Female
    Age group:
      1: Very young (20–40 years of age)
      2: Medium young (41–60 years of age)
      3: Less young (61–80 years of age)
    Order of treatment modality:
      1: CBT for six months followed by medication for six months
      2: Medication for six months followed by CBT for six months

  One within-subjects independent variable:
    Treatment modality
      1: Medication
      2: Cognitive behavioral therapy

Dependent variable: Score on a depression questionnaire
  25 very young male patients with severe depression are treated for 6 months with medication and then with CBT for 6 months.
  25 very young female male patients with severe depression are treated for 6 months with medication and then with CBT for 6 months.
  25 very young male patients with severe depression are treated for 6 months with CBT and then with medication for 6 months.
  25 very young female patients with severe depression are treated for 6 months with CBT and then with medication for 6 months.
  25 medium young male patients with severe depression are treated for 6 months with medication and then with CBT for 6 months.
  25 medium young female male patients with severe depression are treated for 6 months with medication and then with CBT for 6 months.
  25 medium young male patients with severe depression are treated for 6 months with CBT and then with medication for 6 months.
  25 medium young female patients with severe depression are treated for 6 months with CBT and then with medication for 6 months.
  25 less young male patients with severe depression are treated for 6 months with medication and then with CBT for 6 months.
  25 less young female male patients with severe depression are treated for 6 months with medication and then with CBT for 6 months.
  25 less young male patients with severe depression are treated for 6 months with CBT and then with medication for 6 months.
  25 less young female patients with severe depression are treated for 6 months with CBT and then with medication for 6 months.

At the end of each six-month treatment period, a depression questionnaire is completed by the patients.

3.5   Latin Square Designs

A Latin square design is a special type of multifactorial between subjects design that allows the researcher to control for extraneous variables (also known as nuisance variables) by treating them as between-subjects independent variables while managing the number of participants required to complete the study. Remember, the more independent variables a researcher has, the more participants they need in a study. At the same time, there may be factors that could influence the outcome on the dependent variable that are not directly related to the purpose of the research, that is, the main independent variable; thus, the need to factor the effects of those nuisance variables.

In most studies with multiple independent variables, a researcher will have a balanced design. That means that for every level of one independent variable, every level of the next independent variable will be included. So, for instance, in the hand preference, gender, and math aptitude study discussed earlier, there would be two levels of gender for left-handers (i.e. male and female left-handers), two levels of gender for right-handers (i.e. male and female right-handers), and two levels of gender for ambidextrous participants (i.e. males and females who are ambidextrous). This is a balanced design, and is almost always desirable unless one or more of the independent variables is considered a nuisance variable, a variable you don't really care about but need to control for.

This is a little esoteric and abstract, so let's move to a concrete example. In this hypothetical example, we have two nuisance variables: clinic location (four clinics) and provider (four clinicians). We have four different treatments: CBT, Medication, CBT plus Medication, and waitlist control. Our dependent variable is measured after the treatment period; the dependent measure is each participant's score on a depression questionnaire. Treatment lasts 12 weeks. In this example, each patient only participates in one treatment modality. We want to include each treatment in each clinic with each provider to avoid the impact of provider and clinic washing out treatment modality effects – in other words, personal characteristics of the examiner or differences between patients across clinics could be significant factors in determining response to treatment. We don't want clinic location and provider, the extraneous variables, to confound the effects of treatment modality. Therefore, we want to manage or control those variables (location and provider) so that we can detect the unique impact of treatment modality.

If we had a completely balanced design, we would have a $4 \times 4 \times 4$ between-subjects design – four levels of location, four levels of provider, and four levels of treatment modality. If we wanted to have 10 participants in each $4 \times 4 \times 4$ group (i.e. cell), we would have to recruit and treat 640 participants ($4 \times 4 \times 4 \times 10 = 640$). That is, a lot of participants considering we are really only interested in the comparison between four treatment modalities!

So, now let's consider the same example, but using a Latin square design to minimize the number of participants we need to recruit while still controlling for those nuisance variables – location and provider. If we create a 4×4 Latin square (to be illustrated below), we only need $16 \times 10 = 160$ patients. By the way, Latin square designs have to be square; that is, the number of levels in each of the two nuisance variables has to equal the number of levels of the variable of interest – here, we have four levels of provider, four levels of location, and four levels of treatment modality.

The next step is to assign participants to a cell, that is, a group defined by the level of the first nuisance variable, the second nuisance variable, and the independent variable of interest. To do that, we first draw a square and then divide the square into equal parts determined by the number of cells we need. We will have the same number of rows as columns, with the number of each being the same as the number of levels of each independent variable. So, if we have four levels of each independent variable, we will have four columns and four rows, or $4 \times 4 = 16$ cells. Each cell is then filled in with the treatment modality that will be administered to the participants assigned to that cell. The treatment modality is systematically placed in each cell so that each treatment is represented one and only one time in each column (nuisance variable #1) and one and only one time in each row (nuisance variable #2). One method for assigning modality (the main independent variable) to each cell to meet these requirements is illustrated in Figure 3.1. In this example, the final square is labeled as #4.

Now, we assign our participants to groups as illustrated by the cells in the final square. If we want 10 participants in each group, or cell, we will randomly assign 10 participants to be treated by Dr. P at PHS clinic, 10 will be assigned to Dr. P at UT's clinic, and so on, until we reach the final group of 10 participants who will be assigned to Dr. B at SMU's clinic.

#1	Dr. P	Dr. S	Dr. C	Dr. B
PHS	Tx1	Tx2	Tx3	Tx4
UT				
Zale				
SMU				

#2	Dr. P	Dr. S	Dr. C	Dr. B
PHS	Tx1	Tx2	Tx3	Tx4
UT	Tx4	Tx1	Tx2	Tx3
Zale				
SMU				

#3	Dr. P	Dr. S	Dr. C	Dr. B
PHS	Tx1	Tx2	Tx3	Tx4
UT	Tx4	Tx1	Tx2	Tx3
Zale	Tx3	Tx4	Tx1	Tx2
SMU				

#4	Dr. P	Dr. S	Dr. C	Dr. B
PHS	Tx1	Tx2	Tx3	Tx4
UT	Tx4	Tx1	Tx2	Tx3
Zale	Tx3	Tx4	Tx1	Tx2
SMU	Tx2	Tx3	Tx4	Tx1

Note: Tx1 = CBT; Tx2 = Medication; Tx3 = CBT + Medication; Tx4 = Waitlist

Figure 3.1 Sample Latin square design completion.

3.5.1 Summary

While Latin square designs have traditionally been associated with agricultural science, they can be very useful for evaluating intervention studies while managing extraneous nuisance variables as illustrated above. The advantage of Latin square designs is their ability to minimize sample sizes while maintaining control of the extraneous variables and increasing generalizability. Sample sizes and time to complete studies contribute greatly to costs and timeliness in producing and disseminating results, and any strategies for reducing either is invaluable. While some may find Latin square designs out of place in clinical, educational, and behavioral research, it is my opinion that these designs are underutilized due to lack of familiarity.

3.5.2 Double Latin Square Designs

Directly related to the Latin square design is the double Latin square design. A researcher can create a double Latin square to assign participants to the main independent variable (e.g. treatment modality) in a crossover, within-subjects design while controlling for order effects. Remember a crossover within-subjects design is one in which each participant is exposed to all levels of the independent variable, such that each participant is serving as their own control. In the following example (see Figure 3.2), we have three treatment modalities, and every participant receives each and every treatment with each treatment lasting 12 weeks. We vary the order in which different participants will be exposed to each treatment in order to control for carryover effects.

Ultimately, we have six groups, or cohorts, with each group going through the three treatment modalities in one of six possible orders. We have now controlled for order effects, our nuisance variable, without running up the number of participants we need.

3.5.3 Graeco-Latin and Hyper Graeco-Latin Square Designs

Wait, what if we want to use a Latin square design, but we have more than two nuisance variables? All is not lost. It is possible to manage more than two

	Group 1	Group 2	Group 3
First 12 weeks	Exercise	Medication	CBT
Second 12 weeks	Medication	CBT	Exercise
Third 12 weeks	CBT	Exercise	Medication

	Group 4	Group 5	Group 6
First 12 weeks	Exercise	Medication	CBT
Second 12 weeks	CBT	Exercise	Medication
Third 12 weeks	Medication	CBT	Exercise

Figure 3.2 Sample double Latin square design.

nuisance variables at a time using Graeco-Latin and Hyper Graeco-Latin square designs. The Graeco-Latin square is to be used with three nuisance variables, while the even more complex Hyper Graeco-Latin square design is used with four nuisance variables. These designs, reminiscent of Star Trek 3-D chess, are beyond the scope of a primer, but many descriptions can be found online.

3.6 Nesting

Another strategy for managing nuisance variables is nesting. Nesting can be used when it is not feasible to have a completely balanced design, but there is more than one independent variable nonetheless. Nesting seems to be a common strategy when evaluating educational methods. That is because research involving educational methods is often conducted in the classroom setting with classes serving as convenience samples. Unfortunately, convenience samples don't generalize well to larger groups as discussed in Chapter 2. This is, in large part, due to potential systematic differences between one class and another. In order to improve generalizability when employing class samples, a researcher can increase the number of classes that are included in a study; rather than comparing one class to another, they can compare multiple classes receiving an intervention to multiple other classes not receiving the intervention.

To create a fully balanced design when evaluating an education method using classes, a class would have to be divided up, with part of the class receiving one method, and another part of the class receiving another method. However, that might be problematic, for instance, if the method is administered by the teacher of the class. Instead, one teacher could administer one method to their class, and another teacher could administer the alternate method to their class. This is where pre-existing differences between the two classes or between the two teachers can create confounds. The solution is to nest class within method. In other words, assign multiple classes to method #1, and multiple, different classes to method #2. This will allow more generalizability of the findings to a larger portion of the population of interest. Differences between classes receiving the same instructional method can be assessed in order to determine if there are unique contributions of the class versus the instructional method itself. See, *Research Design and Methods: A Process Approach* (Bordens & Abbott, 2011) for a good discussion and illustration of nesting.

3.7 Matching

Matching is another strategy that can be used to manage extraneous variables. Let's say a researcher is interested in the difference between two exercise strategies in managing hemoglobin A1C. In this study, ethnicity may be considered to be an extraneous variable. But, there may not be enough representatives per ethnicity in a given region to obtain enough participants from each ethnic group in order to separately test the effect of ethnicity. Instead, the researcher can at least manage the extraneous effect of ethnicity by matching. That is, the researcher can match the two exercise groups on ethnicity by assigning equal numbers of each ethnic group to each exercise group; for each person of ethnicity "X" in Exercise Strategy #1, the researcher assigns a person of ethnicity "X" to Exercise Strategy #2, and so on. In this way, the exercise groups are at least matched on ethnicity.

Similarly, groups can be matched by using twin studies, with Twin #1 of a sibling pair assigned to one group and Twin #2 assigned to the second group. Twin studies are useful any time a researcher wants to control for the effects of factors such as genetics, family history, socioeconomic status, and/or age.

Even though matching is not quite the same thing as using a within-subjects design, statistically, the data in a matched group design are often treated similarly to within-subjects designs. More specifically, matched group designs are often treated as repeated measures designs. This is because some of the normal variance has been reduced by matching on the extraneous variables, and the groups are not completely independent, or orthogonal, which is a requirement for most parametric statistical tests (to be reviewed in Chapter 5). In this situation, the individuals matched on the extraneous variable are paired, or grouped, with the ultimate comparisons being made across the sets of pairs/ groups.

3.8 Blocking

Blocking is very similar to matching. The term blocking is used to describe what happens when our groups of interest (divided on the basis of the main independent variable) are further subdivided on the basis of a second independent variable. This was described earlier in our discussion of multifactorial between-subjects designs. The second independent variable may also be of interest, or we may consider the second variable extraneous, or nuisance. By subdividing the groups along this extraneous variable, we are controlling for differences associated with that variable.

In a randomized block design, our participants are randomly assigned to one of our experimental groups. But, the participants are first divided into groups on the basis of the quasi-experimental, extraneous variable, and then randomly assigned from their quasi-experimental groups to one of the experimental groups. For example, our participants may first be divided according to gender into a male and a female group. From each of those groups, the participants are then randomly assigned to one of the experimental groups, such as a drug treatment group versus a placebo treatment group.

3.9 Nonexperimental Research

Any of the above study designs can be used when conducting either an experimental, a quasi-experimental, or a nonexperimental study. Whether a study qualifies as being truly experimental in nature depends on whether the criteria for an experimental design are met. Recall, a true experiment requires manipulation (or creation) of the independent variable, random assignment to condition or level of the main independent variable, and control of extraneous variables.

There are certain types of research that are nearly always nonexperimental in nature. Epidemiological studies, for example, represent a form of nonexperimental research in which data are collected either through records or through self-report. Any study looking at the relationship between two naturally occurring factors (e.g. height and foot size) are not truly experimental in nature. Further, studies comparing groups on an innate descriptive characteristic (e.g. gender or hand preference) are not truly experimental. Fixed and innate variables are considered quasi-experimental. This is because variables such as height, foot size, gender, and hand preference are independent variables that are not manipulated by the researcher.

Archival studies, including those relying on an existing database, are not truly experimental in nature. Participants have not been randomly assigned to groups, and extraneous variables have not been controlled. Measurements are rarely obtained by individuals blind to the subjects' conditions, and subjects are rarely blind to treatment modality. Related, there may be very little control over the manner in which dependent variables, such as diagnoses, are measured.

That is not to say that epidemiological or archival studies are bad. On the contrary, valuable information can be gleaned from these generally large-scale studies at a relatively low cost. However, it is very important that researchers do not conclude or convey that their results indicate causality between the independent and dependent variables. There are too many threats to the internal validity of these studies for causality to be deduced.

3.10 Case Studies

As noted in Section 3.9, epidemiological studies and studies relying on large, archival databases are examples of nonexperimental research that can be valuable when conducted and interpreted cautiously. At the extreme opposite end of the nonexperimental research continuum are case studies. In contrast to epidemiological studies, case studies rely on data gathered and observations made with one individual or, in a case series, with a small group of individuals. Case studies can be a useful tool for studying a phenomenon for the purpose of both raising awareness about sometimes rare conditions, and for the purpose of exploring possible theories and developing hypotheses for future research.

Case studies and case series are often presented at conferences, and are sometimes written up for publication with fairly extensive literature reviews concerning the condition under study. These case studies may revolve around an unusual or rare condition, with a description of the course of the condition and the symptoms.

Case studies can also sometimes involve experimentation addressing interventions. In those cases, the experiment is designed in a manner to control for issues such as maturation, history, and practice effects. These designs typically involve obtaining a baseline measure (the dependent variable), implementing an intervention and repeating administration of the dependent measure, removing the intervention and repeating administration of the intervention again, and, finally, reintroducing the intervention and obtaining another measurement of the dependent variable. This particular design is sometimes referred to as an A-B-A-B design, with "A" being the nonintervention condition and "B" representing the intervention condition. An example of an A-B-A-B single-case research design follows in Table 3.7.

Table 3.7 Single-case A-B-A-B research design.

Independent variable: Physical activity during study periods
Levels:
1: No physical activity during the day (8 a.m.–5 p.m.), Monday through Friday
2: 10 minutes of walking outside every 2 hours during the day, Monday through Friday
Dependent variables:
1: Multiple choice practice test covering the week's vocabulary items
2: Mood as measured with a depression questionnaire
The participant is one individual who is preparing for a standardized exam for graduate school applications.
During the first week of the program, the student studies new vocabulary words from 8 a.m. to 5 p.m. each day, Monday through Friday. During the second week of the program, the student studies in the same manner, but adds a 10-minute walk outside every 2 hours. During the third week, the student studies as usual, without the walks. During the fourth week, the student again adds the 10-minute walk outside very 2 hours. At the end of each week, the student completes a multiple-choice vocabulary test that includes all the words reviewed during the week; the four tests (one per week) and the vocabulary words are all of similar difficulty. The student also completes a depression questionnaire at the end of each week.
It is anticipated that vocabulary practice test scores will be higher during each of the two weeks with the walking breaks than during the two weeks without the walking breaks. It is additionally anticipated that depression questionnaire scores will be better (i.e. lower) at the end of each of the weeks that include walking.

3.11   Summary

In Chapters 2 and 3, I reviewed research validity, both internal and external. I also reviewed various research designs. These included experimental designs, with strategies that can be implemented to manage some of the threats to internal and external validity. I additionally briefly reviewed nonexperimental designs. A final type of study I discuss later in this primer is the meta-analytic study. Meta-analyses are studies that systematically review others' research in a technical manner. Meta-analyses depend on a concept called an effect size, which I review in Chapter 4 when I discuss interpretation of statistical analyses.

In the next 3 chapters, I will review inferential statistics. First, in Chapter 4, I will review concepts related to interpretation and application of statistical analyses. In Chapters 5 and 6, I will discuss some of the more basic and prevalent parametric and nonparametric statistics that are used in the behavioral sciences, education, and clinical research. Along with descriptions and explanations of the statistical procedures, I provide examples of studies to which each may be applied. I also briefly summarize some of the statistical software programs that can be used to conduct the analyses.

Section 3

The Nuts and Bolts of Data Analysis

In this section I review major statistical topics. In Chapter 4, as noted earlier, I focus on the interpretation of statistical findings. In Chapters 5 and 6, I provide a cursory review of major statistical techniques. I first review parametric statistical techniques that are commonly used in behavioral and clinical research. In Chapter 6, I review nonparametric statistical techniques that can be used when a study's data do not conform to the requirements of parametric statistics (this will be explained more in the first pages of Chapter 5).

Research Design and Analysis: A Primer for the Non-Statistician, First Edition. Leslie D. Rosenstein.
© 2019 John Wiley & Sons, Inc. Published 2019 by John Wiley & Sons, Inc.

4

Interpretation

After a study has been designed and data are collected, the next step is to analyze the data and then interpret the results. In Chapter 5, I review the different statistical techniques that are available depending on the question being asked and the design of the study. But, first I discuss issues related to interpretation of statistical results that will apply across techniques.

4.1   Probability and Significance

The most common approach at this time for interpreting statistical results is to consider the statistical significance of the findings. Statistical significance is determined by evaluating the likelihood of obtaining a particular value of a statistic. The likelihood of a statistical value is determined by addressing the probability of obtaining that numerical value for a given sample size when there is no effect. For instance, if we are comparing two treatment groups (e.g. placebo versus drug) on an outcome measure, what is the likelihood of obtaining a statistical result (value) if the two groups were equal? If the probability of a statistical result in the face of no difference or no relationship is less than 5 in 100 (0.05), then when we obtain that improbable statistical result we say that the finding is statistically significant at the 0.05 level. In other words, if we obtain a statistical result that is unlikely, or statistically improbable unless there is a true effect (difference or relationship), then we conclude that our result is statistically significant.

By convention, statistical results are often considered statistically significant if the probability of the statistic occurring is less than 0.05. Other than convention, this probability level is somewhat arbitrary, though most would consider the outside 5% of a distribution to be somewhat extreme (i.e. significant). In some studies, the probability level may be set at a different point, say 0.01, depending on the cost associated with decisions made about significance per a

Research Design and Analysis: A Primer for the Non-Statistician, First Edition. Leslie D. Rosenstein.
© 2019 John Wiley & Sons, Inc. Published 2019 by John Wiley & Sons, Inc.

given study. The point at which a statistic becomes significant based on the probability level we have set is called the *critical value* of the statistic.

As a concrete example, let's say that I have conducted a study to compare left-handers and right-handers on a measure of handwriting quality. I have two groups, one group of 15 left-handers, and one group of 15 right-handers. Each participant provides a handwriting sample that is scored on a scale of 0–100. I conduct a *t* test (to be discussed in more detail in Chapter 5), and obtain a resulting *t* of 3.12 (that is, $t = 3.12$). I will look up the probability of obtaining this statistical value ($t = 3.12$) with these sample sizes. According to my table of critical values for *t* scores, the probability of obtaining a *t* that is greater than or equal to 3.12 in this study, with these sample sizes, is less than 0.01 (<1%). Therefore, I conclude that my finding is statistically significant, in this case at the $p < 0.01$ level. If there were no difference in the population between left- and right-handers in handwriting quality, there is less than one in one hundred chance I would have obtained a *t* score of 3.12 with my samples of 15 participants each.

4.2 The Null Hypothesis, Type I (α), and Type II (β) Errors

Certain terms are helpful to know when reading about statistical significance. In particular, the terms null hypothesis (H_0), Type I error (α), and Type II error (β) are critical. The null hypothesis is the hypothesis that there is no effect in our study – such as no correlation between our variables or no difference between the group means. The alternative to the null hypothesis is our actual research hypothesis that our variables are related or our groups are different.

When our statistical value is more extreme than the critical value based on the probability level we have set, we can reject the null hypothesis. That is, we reject the notion that there is no relationship or no difference. We conclude that our result is statistically significant. In the handwriting example, we would conclude that our result is statistically significant, and we reject the null hypothesis that there is no difference in handwriting quality between left- and right-handers.

If we have set our probability level (p) at 0.05, there is a 5% chance that we could be incorrect in rejecting the null hypothesis. This is referred to as a Type I error. Notably, when we set the probability level at 0.05, we often say alpha is set at 0.05 ($\alpha = 0.05$). We are acknowledging that there is a 5% chance that we are incorrect in our conclusion.

On the flip side, if there really is a difference between our groups or there is a significant relationship between our variables, and we do *not* reject our null hypothesis, we have made a Type II error. A Type II error can be thought of as the failure to reject a false null hypothesis. That occurs when our statistical value does not exceed the critical value even when in reality the groups do differ or there is a reliable relationship between variables. There are times when we need to minimize the risk of Type II errors, depending on the decisions that will be made based on our results, and so we increase our α level. An increase in α reduces the critical value of our statistic needed to reject the null hypothesis, which in turn decreases β, the risk of a Type II error.

Finally, it should be noted that when we discuss the probability level for α, we refer to this as our "p value." In papers, authors frequently state something to the effect of "the p value was set at 0.05."

Many people have an easier time grasping the concepts of Type I and Type II errors if they can see them in a table. Table 4.1 presents these concepts, along with the concept of *power* graphically.

4.3 Power

A very important concept in research is power. Power is the probability that you will reject a null hypothesis and conclude significance when there really is an "effect" in the population (see Table 4.1). It is the ability to detect a reliable relationship between variables or a true difference between groups.

Power is impacted by various factors. One of these is the effect size in the population (i.e. the strength of the relationship or the size of the difference). The concept of effect size is reviewed in more detail later in this chapter. For now, the point is that the stronger the effect, or the larger the effect size, the more power you will have. That is because a larger effect is easier to detect.

Another factor that affects power is the α-level. The larger α, the more power you have. That is because you are more likely to reject a null hypothesis with a larger α as discussed above.

Sample size also affects the power of your study. The more participants or subjects (referred to as "N" or "n"), the more power you will have. That is because the larger your sample, the more closely it will reflect the population. Generally, the critical value of statistics needed to reject a null hypothesis will decrease as N increases. Thus, if you have a really large sample size, the critical

Table 4.1 The null hypothesis, power, and errors.

		Truth	
		Null hypothesis is true	Null hypothesis is not true
Our decision	Don't reject the null hypothesis	✓	Type II error (β)
	Reject the null hypothesis	Type I error (α)	Power

value needed to reject the null hypothesis will be relatively small compared to studies with a small sample size.

Let me explain that a little further. When we conduct studies, we use samples. We do not test or treat the entire population to which we are generalizing. If we tested the whole population, then any difference in the mean scores or relationship between two variables would really exist, and we could conclude that we have a real, significant effect. But, when we do research, we use a sample to draw conclusions about the population. As our sample size increases, it approaches the true population size, and the critical value of our statistic becomes smaller.

An additional factor affecting the power of our study relates to the research design itself. Experimental designs in which we have managed some of the error variance due to extraneous factors will have more power than studies with less control. Related, repeated measures (i.e. within-subjects designs) will have more power than between-subjects designs because we have reduced the error variance even further due to participants serving as their own controls. Think about it, in a crossover design, the participants in Treatment #1 are the same as the participants in Treatment #2, so the two treatments are equal on the extraneous subject variables, such as age, gender, intellectual functioning, socioeconomic status, size, and so on.

To understand this better, it helps to have an idea of how statistics are calculated. With statistics such as the F test in analysis of variance (ANOVA), or the t test for comparing group means, the denominator in calculating the statistic is composed of the error variance. That is, the denominator is a factor of the variability in scores that is accounted for by factors other than the experimental factor, the independent variable of interest. By contrast, the numerator is composed of variance *across* experimental groups, the levels of the independent variable. So, any increase in the error variance, the subject variance, will decrease the value of the statistic (see Table 4.2). The smaller the statistic's value, the less likely you are to reject the null hypothesis, and, therefore, the lower your power.

Another strategy for improving power is to compare more extreme groups. Those individuals who differ most on the independent variable of interest will have the most difference on the dependent variable if there truly is a relationship between the independent and dependent variables, barring a curvilinear relationship. To illustrate, let's say a group of researchers hypothesize that parental time spent reading with children impacts reading

Table 4.2 The relationship between error variance and the size of the F statistic.

As the error variance approaches zero, the size of the F statistic increases		
F = Between group variance/error variance	$F = 100/50 = 2$	$F = 100/10 = 10$

performance. The researchers can divide the children into two broad groups based on the amount of time their parents read with them – zero to two hours per week versus two or more hours per week. Alternatively, they may choose to only test the reading performance of those children who fall into two extreme groups based on parent reading time, such as zero to one hour versus more than three hours per week spent reading together. This will increase the likelihood of detecting a significant difference (power) if in fact parent reading time affects the reading performance of children.

Technical issues related to the statistics also impact power, and will be discussed further when the various statistical procedures are reviewed. These include: assumptions, main effects versus interactions, and the use of covariates and blocking. Briefly, for now, violations of assumptions reduce power, main effects have more power than interactions, and using covariates and blocking improves power. Finally, missing data due to attrition or coding errors, for example, will reduce power due to the statistical procedures used to handle those missing data.

4.4 Managing Error Variance to Improve Power

As discussed above, error variance will reduce power, so managing error variance to the extent possible improves power. Error variance can be managed with selection strategies. The more similarity you have between your groups, the less error variance you will have. Groups can be equalized to some extent by using random assignment. Random assignment can be accomplished by using a random numbers generator or a random numbers table. Matching, which was reviewed in Chapter 3, similarly reduces differences between groups on extraneous variables.

Holding extraneous variables constant by only selecting participants in a narrow range on the extraneous variable will also decrease error variance and increase power. For instance, the participants can be selected from a narrow age range. However, as mentioned in the discussion of external validity, this minimizes generalizability; that is, you will only be able to generalize your study's findings to individuals from that narrow age range.

Anytime you reduce measurement error, you will decrease error variance and improve the power of your study. As reviewed in the discussion of internal validity, measurement error can be improved by using reliable and valid measures, ensuring accurate measurement and data entry, controlling the environment so everyone has the same exposure to distraction, and using standardized procedures so everyone in all levels of the independent variable/treatment are processed the same way. Measurement error increases Type II (β) errors because it waters down the association between your independent and dependent variables since the dependent variable is not precisely measured. Remember, a Type II error is the failure to reject a false null hypothesis.

4.5 Power Analyses

Power analyses are used to estimate the power of a study. They allow us to determine the sample size needed to detect a difference given a desired level of confidence and a specific effect size (strength of the effect in the population; a concept that is reviewed in more depth in Section 4.6). Unfortunately, conducting power analyses requires speculating about the true effect size. Funding agencies are often looking for power of 0.8 or 80% (the desired level of confidence). When determining the necessary sample size for a study, one can conduct a power analysis, plugging 0.8 for the value of the desired power into the formula along with the estimated effect size. Effect sizes may be estimated based on prior research in the literature, or theory.

For the true statistics nerd (and no one is judging), there are formulas available for hand calculation of power analyses. However, there are also plenty of programs available for conducting a power analysis to determine the ideal sample size for a planned study. These include, but are not limited to: (1) SAS®'s Proc Power, GLMPOWER, and interactive SAS Power and Sample Size available in Tasks and Utilities; (2) SPSS's Sample Power; (3) Stata's Sampsi; and (4) G*Power (free online for Mac® and Windows®). G*Power is quick and easy to download – you can find the download with a simple online search. Using G*Power is also pretty simple and straightforward.

4.6 Effect Size

When we do research and conduct our analyses, we hope and cross our fingers that our results will be statistically significant. After all, statistical significance supports our hypotheses and increases the chances that our research will be published. But what does statistical significance really tell us? It tells us that we have found a reliable effect that most likely did not occur by chance alone. Statistical significance does not tell us the size or importance of the effect. The effect size of our results tells us about the strength of our effect.

The effect size can be measured in several different ways. How we measure and report the effect size will determine how easy it will be to compare results across studies that have used different measures. You may have read a meta-analysis in the past, and realized that studies are compared using effect sizes. The American Psychological Association (2010) recommends that effect sizes be included in published studies when feasible. This increases the ease with which results from different studies can be compared.

The easiest way to measure an effect size is by reporting the difference in raw scores between groups. Unfortunately, raw score effect sizes may not be useful across studies using different measures or with groups with different variability or distributions of scores (e.g. restricted groups versus broader, more widely representative groups).

Another measure of effect size when we are talking about group differences is the standardized mean difference. The standardized mean difference takes into account that researchers may use slightly different measures to study similar concepts. It also takes into account variance within the samples. Two ways to measure the standardized mean difference are Cohen's d and Hedge's g statistic. Hedge's g statistic corrects for the overestimate produced by Cohen's d in studies with small sample sizes (e.g. $N \leq 20$). It is intended to provide an unbiased estimate of the standard difference in the population. These effect sizes can be calculated by hand or with an online calculator. In either case, you can use the means and variances obtained with stat packages, so you don't have to do any cumbersome calculations by hand.

How does one interpret the numerical value obtained when measuring Cohen's d or Hedge's g statistic? Generally, the researcher should compare and judge their effect size against effect sizes previously obtained and/or published in the literature. The primary value in effect sizes is the comparison of effect sizes across studies. There are published descriptors of effect sizes as being large, medium, or small, but these are somewhat arbitrary and not really meaningful in isolation.

Odds ratios provide an alternative method to talk about effect size. The odds ratio (OR) is the odds of successful outcome in one group (e.g. treatment group) relative to the odds of successful outcome in another group (e.g. controls). When the OR = 1.0, it means there was no effect. Using the OR is the preferred method when outcome data are dichotomous. The OR is relatively easy to calculate by hand.

A similar, but not equivalent measure of effect size with dichotomous outcomes is relative risk. Relative risk is the probability of an event occurring in one group versus another group. The relative risk is typically presented as a multiple, for instance, Group #1 is 2.5 times more likely to succeed than Group #2.

In studies in which the strength of the relationship between two variables is considered, the effect size is measured by the correlation coefficient (r). Taking this a step further, a researcher can use the square of the correlation (r^2) to report the portion of variance in the dependent variable accounted for by the independent variable. So, the correlation statistic tells you the strength of the relationship between two variables, while the square (r^2) tells you how much of the variability in one set of scores (the dependent variable, or criterion) can be attributed to variability in the other set of scores (the independent variable, or predictor). When a researcher is evaluating the relationship between more than two variables at a time, the equivalent effect size measures would be the multiple correlation, R, and R^2.

In Chapter 5, I introduce you to ANOVA. When using ANOVA or one of its related procedures, the effect size can be reported as η^2. The value of η^2, similar to r^2 and R^2, indicates how much of the variance in overall scores is accounted for by group membership. As with R^2 and r^2, the larger the value, the larger the effect of group membership, including with regard to our experimental variable. In other words, say we are looking at depression scores in a medication group versus a placebo-control group. A large η^2 tells us there were big differences in depression scores between the two groups. Related measures when using variations of ANOVA are partial η^2 and ω^2. Thankfully, these effect sizes can easily be obtained when using stat packages such as SAS and SPSS® to conduct ANOVA.

Somewhat related to the concept of effect size is *clinical* or *practical significance*. Remember, statistical significance tells us whether we have a reliable effect, but it does not tell us how big that effect is. The effect size tells us about the strength of the effect. Taking this a step further, we can talk about clinical and practical significance. Practical and clinical significance is the degree to which our findings are important. For instance, in a drug study, we may find a reliable difference in symptom management between a placebo and a drug, but we want to consider the practical significance of the difference. If symptoms are slightly better with the drug, but still intolerable, is it worth using the drug given its side effects? Or, in an intervention study, if a time-consuming and expensive test preparation course reliably improves scores 5 points on a 1600 point college entrance exam, is it worth the time and expense to take the course? Does an increase in five points improve the outcome for the students? The answers to questions about clinical and practical significance are subjective and must take into account the costs and benefits of an outcome.

4.7 Confidence Intervals and Precision

Earlier, I talked about statistical significance. Statistical significance and p values are commonly used today when discussing statistical test results. However, some advocate for the use of confidence intervals, and you will often come across these, sometimes along with statistical significance, in published research. The main drawback of confidence intervals is their seemingly complex nature. At the same time, some argue that they make more sense than statistical significance and p values. In my opinion, confidence intervals can be useful in describing the reliability or importance of findings, but it is most helpful when they are presented verbally and not just numerically. Here, I *try* to briefly describe confidence intervals in a way that makes sense.

A confidence interval, as a measure of precision, provides a range in which the true descriptive statistic (e.g. mean score) of our population of interest falls based on our sample. Say, what? Put another way, the confidence interval is the range of possible values of the descriptive statistic (e.g. mean) we are measuring while taking into account the expected variability caused by sampling error. Sampling error refers to differences that may occur between a particular selected sample and the actual population to which we want to generalize. Smaller samples risk having more error or differences. This is partly due to outliers having a bigger impact in a small sample, and also due to the fact that the larger a sample, the closer it is to actually being the population of interest. If the range of possible values (the confidence interval) of the descriptive statistic that is based on our sample does not include the value of the statistic in the null hypothesis, then we conclude that we have a significant finding. We conclude that our sample's mean is significantly different than the mean in the null hypothesis, and that our sample differs from expectation.

In Table B.5, I provide a sample data set and data analysis using SAS to conduct a t test comparing the IQ test scores of a group of (fictional) swimmers to the general population's mean IQ score, which is 100. The hypothesis of interest is that swimmers have higher IQ scores than the general population. We are testing this hypothesis with a sample of 33 (fictional) swimmers. As can be seen in the example, the group of swimmers are found to have significantly different (in this case, higher) IQ test scores (mean IQ = 117) compared to general expectation (i.e. null hypothesis: mean IQ = 100). The resulting t statistic from the analysis is statistically significant ($t = 10.11$, $p < 0.0001$) meaning that we reject the null hypothesis that our sample of swimmers is not different from the general population (i.e. we accept that they are different). Described another way, the 95% confidence interval around the swimmers' mean IQ ($M = 117$) does not include the general population mean IQ ($\mu = 100$); the 95% confidence limits around our sample's mean range from 113.6 to 120.4. Thus, we conclude that our swimmers' IQ scores were outside the range of general expectation.

As another example, if we are comparing two independent groups on their mean scores, we are interested in the difference between the groups' means. The confidence interval will be the range of possible values of the difference between the means. That difference will be adjusted on the basis of sampling error; smaller samples will have more sampling error, while larger samples, approaching the size of the population, will have less sampling error. If 0 is not in the range of difference scores, the confidence interval, then we conclude that there is a significant difference between the two means. That is because the null hypothesis is that there is no, or 0, difference between the means; if a confidence interval does not include the value of the null hypothesis (0), we reject the null hypothesis that the groups are the same, and conclude that they are different.

If you are confused, you are not alone. While the idea of confidence intervals is sensical at some level, it can be hard to wrap your head around the concept. For a really good description of confidence intervals and how to interpret and report them, I recommend Larry Hatcher's *Advanced Statistics in Research* (Hatcher, 2013).

Finally, as with effect sizes, the American Psychological Association (2010) recommends including confidence intervals for obtained statistics in the *Results* section of published papers whenever feasible. This is because confidence intervals convey added information about results of statistical analyses, including with regard to the precision of sample characteristics.

4.8 Summary

In this chapter, I reviewed concepts necessary for interpreting findings from statistical analyses. These concepts included the null hypothesis, Type I and Type II errors, power, statistical significance, effect sizes, and confidence intervals. When data are analyzed, the resulting statistic must be interpreted. The concepts reviewed in this chapter allow for the interpretation of the statistics in a way that is meaningful. In the next two chapters, I review basic and common statistical procedures.

5

Parametric Statistical Techniques

In this chapter and the following chapter, I review common statistical techniques. This is just a simple listing and description of common statistical procedures available, and when each can or should be used. I do not go into detail about how to conduct the analyses as there are many excellent sources for learning about the specific statistical procedures. If you can find a copy, my all-time favorite for an introductory discussion of basic inferential statistics is an old book coauthored by my first mentor, Dr. Robert Kehoe Young, *Introductory Statistics for the Behavioral Sciences* (Young & Veldman, 1981); otherwise, there are several introductory statistics books available at your favorite online book store or library (a few of which I list in Appendix C). For more advanced topics, it is easy to search online for university course lecture notes on specific topics. I also frequently find helpful information at www.statisticshowto.com (Glen, 2018), https://statistics.laerd.com (Lund & Lund, 2018), www.statisticssolutions.com (Statistics Solutions, 2018), and www.theanalysisfactor.com (The Analysis Factor, 2008–2018). In fact, I relied quite a bit on these sites as I prepared Chapters 5 and 6 of this primer.

5.1   A Little More Lingo

In this chapter, I review some of the more commonly used parametric statistics. Parametric statistics are statistical procedures that can be used to analyze data when certain assumptions are met, and these assumptions generally rely on interval or ratio data that are normally distributed (to be explained further along in this chapter). An assumption is basically a requirement for using a statistical technique. Some assumptions are set in stone, while others can be violated, or breached, to a degree depending on the statistical test and other factors. As I review each statistical procedure, I will discuss the typical assumptions, and I will note when a test is robust to violations of, or deviations from, a particular assumption and under what circumstances.

Research Design and Analysis: A Primer for the Non-Statistician, First Edition. Leslie D. Rosenstein.
© 2019 John Wiley & Sons, Inc. Published 2019 by John Wiley & Sons, Inc.

By the way, a nonparametric statistical test is one that is used when the assumptions of a parametric test cannot be met. So, for instance, a nonparametric test can be used when your dependent variable is measured in terms of frequency or counts (e.g. number of successes, number of participants who survive, number of participants in one category or another). In the case of frequency data, a go-to nonparametric test for looking at differences between groups is the Chi-Square test (to be reviewed in more detail in Chapter 6).

In the following sections, I review some additional lingo needed to understand parametric (and also nonparametric) statistical techniques. Inclusive, I review types of data, central tendency, and distributions.

5.1.1 Population Parameters Versus Sample Statistics

When we conduct statistical analyses, we are typically evaluating samples and making inferences about populations. That is why we call our type of research statistics, "inferential statistics." We often do not know the exact descriptive statistics, or characteristics, of a population, but we can measure those characteristics in our sample and then make inferences about the population. As our sample size becomes larger, our sample statistics will approach, or become more similar to, the population parameters.

When discussing statistics using symbols and letters, you will note that population parameters are typically denoted with the Greek alphabet, while sample statistics are labeled with Latin letters. For instance, the population mean is typically denoted as μ while the sample mean is indicated by M or $\bar{X}$.

5.1.2 Data

One of the assumptions that will be consistent across the parametric statistics that I review in this chapter relates to the type of data that can be analyzed. Generally speaking, data analyzed with parametric statistics will be interval or ratio in nature as mentioned earlier.

5.1.2.1 Ratio and Interval Data

Ratio data are numbers that have an absolute zero value, and an equal distance between each number and the next such that a data point can be expressed as a ratio. Interval data meet the same criteria as ratio data, except that there is no absolute zero value, so ratios cannot be calculated to reflect fractions. However, the distance between two consecutive numbers is always the same. Interval and ratio data are both considered to represent continuous variables.

5.1.2.2 Ordinal Data

A step down from interval data is the concept of ordinal data. As can be imagined, ordinal data have order that has meaning. Ordinal data can be thought of as ranks. The ranks, though, are not equidistant from one another in terms of the quantity of the underlying variable being ranked. For instance, if the ordinal variable is rank order of height in a group (from tallest to shortest), the difference in height (the underlying variable) between the tallest and second tallest person may not be equal to the distance between, for example, the 10th and 11th tallest people in the group.

5.1.2.3 Nominal Data

Finally, one more step down is the concept of nominal data. Nominal variables are not numerical, though they can be assigned dummy codes in the form of numbers, but the numbers do not have a mathematical value. They are simply categories or names. We often refer to nominal variables as frequency data because we use counts of the number of members in a particular nominal group (e.g. 50 females and 40 males, 20 left-handers and 70 right-handers). The categorical variables in this example are gender and hand preference, and the frequency data are reflected in the numbers of members in each respective group.

5.1.3 Central Tendency

Central tendency refers to the center of a range of values or scores. The major types of central tendency utilized in statistical analyses are mode, median, and mean. You will see in subsequent sections and also in Chapter 6 that the mean and median, in particular, play a central role (no pun intended) in the calculation of many inferential statistics.

5.1.3.1 Mode

When we talk about the mode, we are referring to the most common value among a group of discrete values. By discrete, I mean that the values are limited or finite, such as 1, 2, 3, 4, etc., as opposed to continuous. A continuous variable has an infinite number of values, such as 1.135. 1.136. 1.17, 1.9, 2, and everything in between. The mode then is the most common score from a set of discrete scores. For instance, if we have 15 participants in our study, and we score them on a measure that ranges from 1 to 10, the mode is the most common score. If four of the participants score 5, three attain a score of 6, two attain a score of 4, two score 8, two score 3, one attains a score of 2, and one attains a score of 9, the modal score, or mode, is 5. That is because more participants obtained a score of 5 than any other score.

5.1.3.2 Median

The median, on the other hand, is grossly the middle score among a range of scores that are rank ordered from the lowest to the highest score. The median is not necessarily half the value of the highest score. Rather, it is the score that falls midway in the ranked scores. This is illustrated in Table 5.1.

Table 5.1 Fictional illustration of mode, median, and mean.

Independent variable: Athlete type
  Levels: 3
    1: Baseball player
    2: Lacrosse player
    3: Swimmer

Dependent variable: Number of meals consumed per week

Sample: All athletes at a large Division 1 university were included in the study:
  35 are members of the baseball team
  46 are members of the men's Lacrosse team
  34 are members of the men's swimming team

Each participant kept a daily diary of the number of meals they consumed for a week. An athletic trainer who is studying kinesiology totaled each athlete's number of meals for the week and recorded those numbers into a data set (see below).

Data (Meals per week per athlete):

Baseball	Lacrosse	Swimming
21	28	49
7	32	42
14	26	33
42	28	42
21	33	35
39	30	46
42	32	48
35	28	42
14	9	33
7	40	47
21	14	42
17	42	34
15	25	45
34	28	48
35	21	42
21	42	35

Table 5.1 (Continued)

11	24	49
14	35	44
21	11	35
35	14	42
35	28	35
40	21	44
42	19	35
14	14	42
28	35	30
28	27	43
30	21	28
21	28	39
21	29	42
17	31	43
18	35	40
18	26	40
16	28	42
14	21	43
21	14	
	21	
	35	
	28	
	21	
	14	
	28	
	31	
	25	
	28	
	14	
	22	

(Continued)

Table 5.1 (Continued)

The athletic trainer/researcher then calculated the frequency of each score for each team (see below):

Frequency of each score

Meals per week	Baseball	Lacrosse	Swimming
7	2		
9		1	
10			
11	1	1	
14	5	6	
15	1		
16	1		
17	2		
18	2		
19		1	
21	8	6	
22		1	
24		1	
25		2	
26		2	
27		1	
28	2	10	1
29		1	
30	1	1	1
31		2	
32		2	
33		1	2
34	1		1
35	4	4	5
39	1		1
40	1	1	2
41			
42	3	2	9
43			3
44			2

Table 5.1 (Continued)

45			1
46			1
47			1
48			2
49			2

Mode: From the frequency table above, we can see that the modal number of meals consumed per week among the baseball players was 21 ($N = 8$). The mode for the lacrosse players was 28 ($N = 10$). Finally, the mode of the swimmers was 42 ($N = 9$). That is, the most common number of meals among the baseball players was 21 per week, with 8 players reporting that frequency. Among lacrosse players, 9 reported consuming 28 meals per week, more than any other frequency for that group. Among swimmers, the most common frequency was 42 per week, with a total of 9 swimmers reporting consumption of 42 meals per week.

Median: To help clarify the calculation of the median to the coaches and dieticians for each team, the athletic trainer organized the number of meals per athlete in rank order (see below):

Rank-ordered # Number of meals per athlete

Rank	Baseball	Lacrosse	Swimming
#1	7	9	28
#2	7	11	30
#3	11	14	33
#4	14	14	33
#5	14	14	34
#6	14	14	35
#7	14	14	35
#8	14	14	35
#9	15	19	35
#10	16	21	35
#11	17	21	39
#12	17	21	40
#13	18	21	40
#14	18	21	42
#15	21	21	42
#16	21	22	42

(Continued)

Table 5.1 (Continued)

#17	21	24	42
#18	21	25	42
#19	21	25	42
#20	21	26	42
#21	21	26	42
#22	21	27	42
#23	28	28	43
#24	28	28	43
#25	30	28	43
#26	34	28	44
#27	35	28	44
#28	35	28	45
#29	35	28	46
#30	35	28	47
#31	39	28	48
#32	40	28	48
#33	42	29	49
#34	42	30	49
#35	42	31	
#36		31	
#37		32	
#38		32	
#39		33	
#40		35	
#41		35	
#42		35	
#43		35	
#44		40	
#45		42	
#46		42	

Using the rank-ordered table of scores above, we can calculate the point at which half the scores fall above and half fall below for each group. That will be the median.

Table 5.1 (Continued)

Baseball players: Having rank-ordered each and every player using the frequency table above, we see that the middle player (#18 out of 35) consumed 21 meals per week (along with 7 of his teammates). Thus, the median number of meals consumed among the baseball players was 21 – same as the mode for that group.

Lacrosse players: There were 46 lacrosse players, so the median score will be the score that falls midway between the 23rd ranked player (in terms of meals consumed) and the 24th player. As it turns out, we again have a lot of ties in the middle, such that the 23rd and the 24th ranked number of meals was 28 (as was the number of meals for 8 of the other players). So, the median is 28, again the same as the mode.

Swimmers: Since there were 34 swimmers, the median will equal the midpoint between the number of meals consumed by the 17th and 18th ranked eaters among the swimmers. Again, the mode equals the median at 42.

Mean: Finally, we can calculate the mean by summing all of the scores for each group, and dividing each sum by the respective number of athletes in each group.

Baseball: So, using the original data set, we find that the sum of the number of meals among baseball players for the week was 829. Diving that number by 35, the total number of baseball players, we see that the average number of meals per week is 829/35 = 23.69. That is, on average, the baseball players consumed 23.69 meals per week.

Lacrosse: The sum of the number of meals consumed by the lacrosse players during the week was 1186. Dividing that sum by the number of lacrosse players, 46, gives us 1186/46 = 25.78.

Swimmers: The sum of all meals among swimmers was 1379, so Mean = 1379/34 = 40.56.

Sometimes, there are a lot of ties in the score in the middle of the distribution. In that case, the median will be the score that is tied with the other scores. Often, the number of scores falling above and below the median will not be the same – that occurs when there are tied scores, and the ties are not precisely in the middle of the distribution (see Table 5.1, "Median" of the Lacrosse players for a good example of this). When the number of scores in a range of scores is even, the median will be the number that is midway between the two middle scores. So, if you have 12 scores, the median will be the score midway between #6 and #7 when the scores are ranked in order. Another way to think about that is that the median will be the average of those two scores (i.e. the average of the score corresponding to the #6 ranked observation, and the score corresponding to the #7 ranked observation).

5.1.3.3 Mean

Another common term for the mean is the average. As you may already know, the average is calculated by totaling all the scores, and dividing that sum by the number of scores. This is often shown as $\Sigma X/N$ – the sum of the X scores divided by the number of X scores in the sample.

In Table 5.1, I present a set of discrete data from a fictional study of the number of meals consumed per day among various types of athletes. From these data, you can see the mode, the median, and the mean.

Now that we have reviewed means, medians, and modes, we can review distributions. Specifically, we will review frequency distributions, that is, distributions of scores.

5.1.4 Distributions

When we talk about distributions of scores in inferential statistics, we are talking about the layout of scores by frequency. So, for instance, in the bell-shaped, normal curve, the top of the distribution is the point, or score for which there is the highest frequency. Even though the vertical axis is not always labeled, the normal distribution is really a frequency distribution with frequency indicated by the vertical axis as illustrated in Figure 5.1. As can be seen in the distribution of scores in Table 5.2, which happen to be normally distributed, the mean, median, and mode are equal. As we know from the discussion in the previous section, the mode is the score with the highest frequency of occurrence. And, in a normal distribution, which will be symmetric, the mode, median, and mean will be about equal. In addition, in a normal distribution, there will be a fairly significant drop off in frequencies as we move away from the mode, with very few scores occurring at the extremes.

The distribution in Figure 5.1 is contrived from the data in Table 5.2.

Not all distributions are normal. That is, not all scores are normally distributed. That would make things too easy and boring! Distributions can be flat, they can be tall and skinny, they can be skewed toward one side or the other, and they can have more than one mode (as in a bimodal distribution).

A tall and skinny distribution occurs when there are a few outliers in the extremes, but with more scores found in the middle. This is considered a

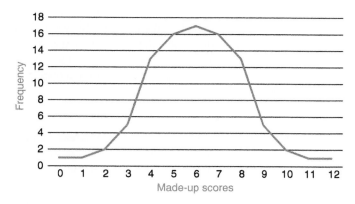

Figure 5.1 A roughly normal distribution.

Table 5.2 Data illustrated in the normal distribution in Figure 5.1.

Score on made-up test	Frequency of occurrence
0	1
1	1
2	2
3	5
4	13
5	16
6	17
7	16
8	13
9	5
10	2
11	1
12	1

positively kurtotic distribution, also described as leptokurtic (see Figure 5.2). A negatively kurtotic distribution will appear squattier (platykurtic) due to having more scores in the tails, and not as much concentration in the middle; the tails are not composed of extreme values or outliers (see Figure 5.3). Visually, a highly kurtotic distribution looks like it has a higher peak, while a negatively kurtotic distribution looks like it has a lower peak; however, this is due to the relative concentration at the tails versus the center (Tables 5.3 and 5.4).

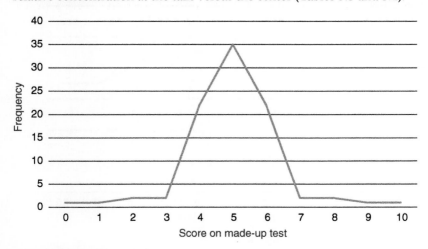

Figure 5.2 Positively kurtotic distribution.

Table 5.3 Data illustrated in the positively kurtotic distribution in Figure 5.2.

Score on made-up test	Frequency of occurrence
0	1
1	1
2	2
3	2
4	22
5	35
6	22
7	2
8	2
9	1
10	1

Figure 5.3 illustrating a negatively kurtotic distribution is based on the frequency data in Table 5.4.

While the distributions discussed so far are symmetric around the center, distributions can also be skewed. They can be skewed positively or negatively. In the case of a positively skewed distribution, the tail is longer on the right. A negatively skewed distribution has a longer tail on the left. These are illustrated

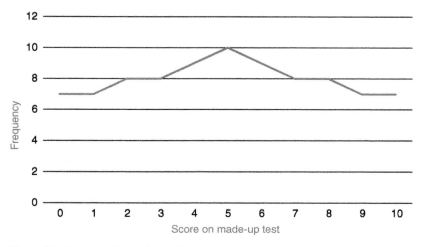

Figure 5.3 Negatively kurtotic distribution.

Table 5.4 Data illustrated in the negatively kurtotic in Figure 5.3.

Score on made-up test	Frequency of occurrence
0	7
1	7
2	8
3	8
4	9
5	10
6	9
7	8
8	8
9	7
10	7

Figure 5.4 A positively skewed distribution.

in Figures 5.4 and 5.5. When a distribution is skewed, the mean, median, and mode will no longer be the same. As illustrated by Stephanie Glen on the Statistics How To website (Glen, 2018), in a negatively skewed distribution, the mean will be pulled down, toward the tail, such that it will be smaller than the mode. The median will be somewhere between the mode and the mean. In a positively skewed distribution, the mean will be pulled up toward the tail, such that the value of the mean is higher than the mode, again with the median midway between the two.

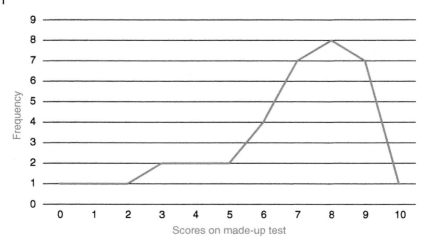

Figure 5.5 A negatively skewed distribution.

When a distribution has more than one mode, we can refer to it as multi-modal. If there are two modes, it is called a bimodal distribution. A bimodal distribution occurs not infrequently in the real world, typically indicating that our sample or population is really composed of more than one group – each mode reflects a different group. A bimodal distribution does not need much explanation, but I have included an illustration for good measure (see Figure 5.6).

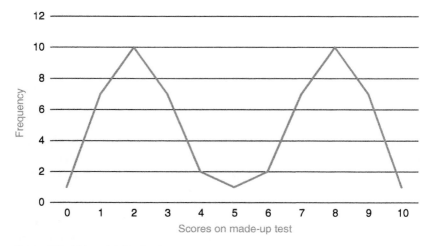

Figure 5.6 A bimodal distribution.

So far, we have reviewed distributions in terms of their mode, median, and mean as well as their shape. Another important aspect of distributions is the spread. The spread gives us an indication of the expected or observed range of scores. Three especially important concepts in spread are *range, variance,* and *standard deviation.* The range is simply the distance between the high and low values in the distribution (found by subtracting the lowest value from the highest value).

Variance is a slightly more complicated way to depict the spread of scores, and is based on the average squared deviation of scores around the mean. Literally, it is calculated by summing the squared deviations of each score from the mean, and then dividing that sum by a factor of the sample size (e.g. in the case of samples that we use to estimate population parameters, we use $N - 1$: sample size minus 1). Notice that a range can be heavily influenced by the presence of an outlier at either end of the distribution, or both. So, for instance, most of the scores in a distribution of scores may fall between, say, 20 and 80, but an outlier may be present at each end, say at the minimum and maximum scores of 0 and 100. In the former case, the range is only 60, while the range with the outliers present is 100. For this reason, it is beneficial to have other measures of spread that can be used in statistical analyses when we want to make generalizations from our sample to the population. The variance and its relative, the standard deviation, serve this role.

Many statistical analyses use the variance or a derivative of the variance in their calculations. The derivative is actually the square root of the variance, and it is called the standard deviation. As the square root of the variance, the size of the standard deviation is in line with the original unit (measurement) size. As stated, the variance and the standard deviation are utilized in the calculation of many different statistics. In fact, the importance of variance, as in error variance, is something we reviewed in the discussion on power. The concept of variance will be highlighted again in the discussion of analysis of variance (ANOVA) further along in this chapter, where we will discuss the variance between groups relative to the variance in the error term.

When we conduct statistical analyses, we are often interested in testing for differences between groups. When we are testing for differences, we are not just testing the difference between the exact mean or median of two (or more) sets of scores; rather, we are testing for differences between the distributions of the two sets of scores. This is one reason why many of our parametric statistics require normal distributions with approximately equal variances (to be reviewed more as each statistic is reviewed later in this chapter). If the distributions have a lot of overlap, they are not really different. Of course, with very large samples, any deviation from perfect overlap could indicate a reliable, statistically significant, though perhaps small, difference. If there is little overlap between our distributions, we are looking at differences between the groups. These concepts are illustrated in Figures 5.7 and 5.8.

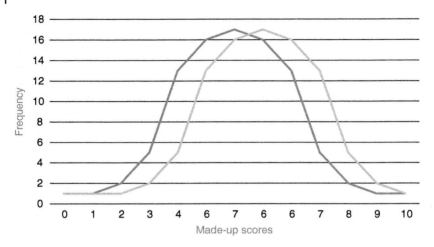

Figure 5.7 Distributions with substantial overlap.

Figure 5.8 Distributions with little overlap.

5.1.5 Dependent Variables

Before we get into the nitty gritty of specific statistical analyses, it is time to discuss the dependent variables. We have already reviewed the concepts of nominal, ordinal, interval, and ratio data. The next consideration with respect to our dependent variable pertains to the choice of whether to use raw scores, percentile ranks, z-scores, other standardized scores, or transformed scores. These are all choices that will apply to interval and ratio data.

Raw scores are scores as reported in their original metric. For instance, total pounds or total inches are raw scores when referring to weight and height. Similarly, if I have a test with 100 items, the raw score will be the number of correct items. Raw scores provide the most precise detail and discrimination between individual observations, but they can have significant disadvantages.

First, a raw score will be meaningless to anyone who is not familiar with the measure being used. Most people have more than adequate familiarity with pounds and inches (sadly). In contrast, a raw score of 49 on a hypothetical test will mean little to most individuals without knowledge or experience using that instrument. A raw score of 49 means nothing out of context and without information about typical scores. If the scores tend to range from 25 to 50, then one could potentially conclude that 49 is a high score on whatever construct is being measured. But, interpreting a score of 40 might be more difficult to interpret out of context without knowing the typical variance or the shape of the distribution of scores. There may be very few scores over 35, for instance, making a score of 40 pretty high. Without knowing the details of the distribution and without familiarity with the test, a raw score will not be meaningful to most people.

Some individuals with a grasp of math will be able to comprehend percentile ranks in lieu of raw scores regardless of their familiarity with a test or measure. Even height can be reported as a percentile, and parents and pediatricians often rely on percentile ranks when interpreting the height of a child. Is 41″ a normal height, for example, for a 5-year-old girl in the United States? Pediatricians routinely track a child's growth on a chart in which height is converted to a percentile rank. In fact, looking at the growth chart produced by the National Center for Health Statistics (2000), a height of 41″ places a 5-year-old girl at the 25th percentile, meaning she is taller than 25% of the population in her age group and gender, and shorter than 75%.

Another disadvantage of raw scores is their potential lack of equivalency across time, samples, and administrations. The Law School Admissions Test (LSAT) offers a perfect example of this situation. The LSAT items are changed from administration to administration so that examinees cannot learn of the test items in advance. But that means the test will not be exactly the same from administration to administration and, thus, the raw scores will not have exactly the same meaning. For this reason, the Law School Admission Council (LSAC) standardizes the scores to a scale with a range of 120–180. Additionally, from year to year, a score attained within the range will convert to approximately the same percentile rank. For instance, a scaled score of 159 indicates a percentile rank of 77.6, meaning that a scaled score of 159 in 2011 was higher than 77.6% of examinees' scaled score during the three previous years (alphascore, 2015). The important piece in terms of the equating that is performed is that it makes the scaled scores of applicants comparable from year to year in spite of the test's content which must change for every administration.

Intellectual Quotient (IQ) scores are another prime example of a type of test score that has been converted from raw scores to a score that is scaled according to the normative sample's characteristics. Specifically, most IQ test scores are standardized to have a mean of 100 and a standard deviation of 15. That means that an average IQ score will be 100, while an IQ score of 85 will be one standard deviation below the mean, and an IQ score of 115 will be 1 standard deviation above the mean.

As it turns out, a score can also be converted to a percentile rank according to the number of standard deviations it falls from the mean if the distribution of scores is normally distributed in the population. If you use statistics in your everyday work life, you will have memorized that a score corresponding to one standard deviation below the mean (like an IQ score of 85; 100−15 = 85) is the equivalent of a percentile rank of 16th, while a score corresponding to one standard deviation above the mean (like an IQ score of 115; 100+15 = 115) indicates a percentile rank of 84th. Now, if you don't have the normal distribution percentile ranks floating around freely in your cortex, then you can always refer to a set of tables in the back of a statistics book or online by searching for, "standard normal distribution."

This discussion brings me to the phenomenon of z-scores. A z-score is simply the number of standard deviations that a score falls from the mean for the population when the population distribution is normal. A z-score can be calculated anytime you know the population mean and standard deviation. For instance, if you have a raw score on a test, and you know the population mean and standard deviation of the test, then you can calculate the number of standard deviations that the score falls from the mean. z-scores can be positive or negative, depending on whether the corresponding raw score falls above or below the mean, and z-scores can include fractions (typically expressed with decimal places). As indicated in the discussion on IQ scores, a z-score of −1.00, which is one standard deviation below the mean, corresponds to a percentile rank of 16th, while a z-score of +1.00 will correspond to a percentile rank of 84th.

Since in statistics we are often striving for statistical significance defined as something occurring less than 5% of the time ($p < 0.05$), it is helpful to know the positive and negative z-scores that correspond to percentile ranks of 2.5 percentile (since $p < 0.05$ really means the sum of the 2½% at each end of the distribution). The z-score that corresponds to the point above which 2.5% of scores fall in a normal distribution is 1.96. Thus, since the normal distribution is symmetric, we also know that the z-score that corresponds to the point below which 2.5% of scores fall (at the other end of the distribution) is −1.96. Don't worry, if you use statistics on a regular basis, you will probably be able to recite $z = 1.96$ in your sleep after a while. If not, you can always look in a z-score table online or in the back of a statistics text book. The main point is that scores that convert to a z-score of ±1.96 will be considered statistically significant because they deviate significantly from the mean for the population.

Two additional types of standard scores you may encounter are T scores (not to be confused with t tests) and Stanines. These are standard scores used in educational and psychological testing. A T score is a score that has been converted from raw scores, with the distribution of T scores having a mean of 50 and a standard deviation of 10. Stanines have a mean of 5 and a standard deviation of 2.

Finally, let's talk really briefly about score transformations when scores are not normally distributed to begin with. It is possible to transform scores using mathematical procedures in order to create a distribution that is normal. You might want to do this in order to be able to use parametric statistical tests. There are several different procedures that can be applied to each score to create a normal distribution, though three common ones are the log transformation, the square root transformation, and the arcsine transformation, all reviewed succinctly in the online *Handbook of Biological Statistics* (McDonald, 2014) as well as plenty of other online sources. Dr. McDonald even provides the statements you would use in the Data step if you want to transform your data for analysis with the SAS® statistical software package.

5.1.5.1 To Scale or Not to Scale

With some derived scores (e.g. scaled scores developed through inferential norming), you may lose some of the detail that you have in raw scores. Loss of detail may mask group differences (see the example in Table 5.5). Therefore, it is critical that you understand how the derived scores are developed. For instance, some scaled scores are derived using a method called inferential norming as indicated above (see Wechsler, 2008a; Zhu & Chen, 2011). While important for certain tasks (such as developing normative data), this type of score transformation may reduce the detail in the scores by condensing sets of raw scores into scaled scores. In other words, a range of raw scores are made to equate to the same scaled score. Thus, you have lost the ability to distinguish between those individual scores within a scaled score. Directly related, two scores with a certain amount of spread between them may equate to a particular scaled score, while two scores with less spread between them may convert to two different scaled scores. For instance, let's say that raw scores of 1 through 5 on a test equate to a scaled score of 1, and raw scores of 6 through 10 equate to a scaled score of 2. As you can see, the highest raw score with a scaled score derivation of 1 is closer to the lowest raw score with a derivation of 2 than it is to the lowest raw score with a derivation of a scaled score of 1. That is, a raw score of 1 equates to a scaled score of 1 as does a raw score of 5. Just one point up on the raw score scale, a raw score of 6 jumps to a scaled score of 2. Hopefully, this will become clearer as you review the example in Table 5.5.

In Table 5.5, I demonstrate a dependent t test (which we will review in more detail in Section 5.2) comparing Wechsler Adult Intelligence Scale-IV (WAIS-IV; Wechsler, 2008b) Block Design subtest performance among a small sample

Table 5.5 Dependent *t* test using raw and scaled scores.

Independent variable: Vision correction
  Levels: 2
    With glasses
    Without glasses

Dependent variable: WAIS-IV Block Design performance

Data set:

Scores on the WAIS-IV Block Design subtest

Raw scores				Scaled scores		
No glasses	Glasses	Difference		No glasses	Glasses	Difference
0	5	5		1	1	0
9	13	4		3	3	0
9	13	4		3	3	0
35	38	3		8	8	0
35	38	3		8	8	0
35	38	3		8	8	0
35	38	3		8	8	0
53	55	2		12	12	0
53	55	2		12	12	0
61	62	1		15	15	0

Raw score $t_{(9)} = -8.22, p < 0.0001$.   Scale score t = unattainable, 0 difference.
When using the raw scores, we see that there is a significant improvement in Block Design test scores when the participants wore glasses; when we converted the raw scores to scaled scores using the WAIS-IV Conversion Tables, the improvement is masked by loss of detail, or discriminability, in the scores.

of participants with and without their glasses. In this example, I first calculate the dependent *t* test using raw scores for the Block Design subtest. I then attempt to calculate the dependent *t* test after converting the raw scores to scaled scores. As can be seen, this conversion to scaled scores results in a masking of the differences by reducing the discriminability of the scores. In this extreme example (which I manipulated for the sake of demonstration), the result is the appearance of no change in performance when using scaled scores, whereas the raw scores demonstrate a clear difference between the scores among participants with and without glasses.

5.1.6 Summary

In this chapter, I reviewed concepts relevant to inferential statistics, particularly with regard to central tendency, spread, and types of scores. Now that you are completely familiar and conversant with the statistics lingo and concepts, we can move on to discussing different inferential statistical procedures. Again, this is a cursory review of some of the major techniques, and does not include a lot of detail regarding the actual mathematics involved. See Appendix C for a list of recommended textbooks and websites for learning more on the how-tos of statistics as well as discussions of more advanced statistic topics.

5.2   t Tests

No doubt, if you have ever taken an inferential statistics course, you remember something about t tests. In a nut shell, t tests are used when you want to know whether two means are different. You can compare two sample means or a sample mean to the theoretical or known population mean (μ) by using a t test for independent samples. You can also assess the difference between pairs of scores with the matched pairs t test. The matched pairs t test is also referred to as a dependent t test.

5.2.1   Independent Samples t Tests

With the independent samples t test, a researcher can compare two groups on a dependent variable. For instance, a sample of Kindergarten girls can be compared to a sample of Kindergarten boys to evaluate whether their heights differ at that age. Basically, if the difference between the two sample mean heights is significantly larger than expectation based on the variance of the heights and sampling error, the difference between the groups is considered significant. In case anyone ever asks, and they won't, the value of t when comparing two samples is the difference between the means divided by the square root of the sum of the squares of the two standard errors of the mean.

The larger the value of t, the more likely it will exceed the critical value for a given sample size in order for the null hypothesis to be rejected. In other words, statistical significance will be concluded if t is larger than the critical value for a given sample size. Most computer statistics packages now include the exact p value for an obtained value of t, though the critical values of t at the 0.05 and 0.01 levels can generally be found in any statistics text book. The critical values for each level are listed for various sample sizes under the heading "df" for degrees of freedom. When comparing two samples, the degrees of freedom used will be the sum of the sample size of the first group minus 1 and the sample size of the second group minus 1 (or, the sum of the combined sample sizes minus 2).

An independent t test can also be obtained when comparing a sample mean on a measure to the predicted mean in the population. For instance, going back to the height example, a researcher may be interested in knowing whether a sample of Kindergartners who have been living on an outpost on Mars where the gravity is below that of Earth are taller than their Earthbound peers. The mean height of the sample of Kindergartners on Mars can be compared to the generally accepted average height of Earthbound Kindergartners.

In this case, the researcher measures the height of the sample from Mars, and compares that mean (say 43.5″), to the theoretical population mean (μ) height of the typical Earthling Kindergartners (say, 42″). Here, the value of t is the difference between the sample mean and μ, divided by the standard error of the mean. The standard error of the mean is the expected variability of sample means around the true population mean, with more variability expected for smaller samples, and less variability expected for larger samples; it is calculated as the square root of the sample variance divided by the square root of the sample size. If the standard error of the mean is 0.5 for our sample of 25 students living on Mars, then $t = (43.5 - 42)/0.5 = 3$. In this case, our degrees of freedom is 24 since degrees of freedom in a single sample t test is $N - 1$. The critical value of t for statistical significance with α set at 0.05 and degrees of freedom = 24 is 2.06. Since our obtained value of t (i.e. 3) is larger than 2.06, we conclude that we have a statistically significant result, and we reject the null hypothesis. That is to say, we reject the notion that our students on Mars are the expected height for Earthlings in their age range, and we conclude that this sample from Mars is statistically significantly taller than typical for human children.

5.2.2 Matched Group Comparison

In Chapter 3, I noted that using participants as their own controls, as in a repeated measures design, can help manage error variance that is due to extraneous variables. I also mentioned that pairs of participants, like twins, can be employed in studies to help control some of that error variance caused by groups differing on more than just the variable of interest. When comparing scores obtained from the same individual at two points, or from each member of a pair, the proper t test to compute and interpret is the matched pairs t test, also known as a dependent t test.

It is critical when comparing means to use the correct t test – either the matched pairs t test or the independent samples t test – because they are very different. Conceptually, they are the same in that two groups of scores are being compared, but the calculation and the critical values of t are different.

In the t test for matched pairs, the numerator of the t test is the average difference between each pair of scores, and the denominator is the standard error of the difference scores. The degrees of freedom for determining the critical value of t is the number of pairs minus 1. The reason for using the different formula is that the two sets of scores are related, and they will potentially be correlated to some extent due to that fact. Therefore, you would expect there to be less difference between the mean scores, and so the denominator (the error term) should take that into account.

Interestingly, at a mathematical level, the t test for matched pairs is computationally the same as the single sample t test, except you use difference scores in lieu of individual scores. In the case of the matched pairs t test, the mean (or average) difference score is compared to a population mean (μ) of 0. A statistically significant t value is then interpreted as meaning that there was a significant difference in pairs of scores, meaning each participant's or pair's first score differed significantly from their second score on average; that is, the difference between paired scores was significantly larger than 0.

Let's consider an illustration of a study for which the paired t test would be appropriate. Let's say we administer a vision test to individuals with and without correction for myopia. We control for practice effects by using alternate versions of our vision test. One hundred participants take the vision test without their glasses and again (alternative version) with new glasses. To be sure that any difference is not due to the version of the vision test, half the participants get Version A of the vision test first, and half get Version B first, and vice versa. To calculate the dependent t test, or matched pairs t test, we subtract the second measurement from the first measurement for each participant to determine their difference scores. We then calculate the average of the difference scores (the mean difference), and divide that number by the standard error of the difference. A significant t tells us that there was a significant difference between the two sets of measures across our participants. Assuming that the vision scores were higher with correction, we would conclude that vision scores were significantly higher when participants used correction.

5.2.3 Assumptions of t Tests

The t test is used when you have interval data or ratio data – that is, when you have a variable such as height, age, weight, score on a 100-point test, and so forth. Scores should also come from a relatively normal distribution (i.e. a set of scores with a mean somewhere in the middle, with few scores at the extremes – often characterized as a bell-shaped curve). It is also assumed in the independent t test that your participants have been randomly selected so that they are representative of the population. Finally, the variances (spread) of the two samples of scores should be approximately equal – this is called homogeneity of variance.

5.2.4 More Examples of Studies Employing *t* Tests

As reviewed earlier, *t* tests can be used to compare either two independent groups, or two sets of scores from the same group of individuals, or from pairs of related individuals. Examples of each are described in Tables 5.6 and 5.7, and a single sample *t* test is included in Table B.5.

Table 5.6 Example of a *t* test for independent groups.

Independent variable: Treatment
  Levels:
    1: New adjunct medication
    2: Care as usual
 	
Dependent variable: Beck Depression Inventory-II Score
 	
Sample: All participants are recruited from an outpatient psychiatry clinic where they have been diagnosed with major depression:
  50 are treated with their care as usual
  50 are treated with the addition of a new medication
 	
Each patient is randomly assigned to either the control (care as usual) group or the new adjunct medication group after consenting to be in the study. After a six-month treatment period, patients complete the Beck Depression Inventory-II.
 	
An Independent Groups *t* test is conducted to compare Beck Depression Inventory-II scores between the two groups to see if those receiving the new medication report less severe depression than the control group.

Table 5.7 Example of a *t* test for matched pairs.

Independent variable: Vision correction
  Levels:
    1: Without glasses
    2: With glasses
 	
Dependent variable: Performance on a visual scanning speed task
 	
Sample: 20 participants with low vision.
 	
Each participant completes a visual scanning speed task with glasses (the Correction Condition) and without glasses (the No Correction Condition). Half the participants complete testing with correction first and then without, and the other half complete testing without and then with correction.
 	
A matched pairs *t* test is used to compare test performance in the No Correction condition to the Correction condition on the visual scanning task to determine whether Correction was associated with improved performance.

5.2.5 Statistical Software Packages for Conducting *t* Tests

In Appendix B, I include a listing and description of various statistical packages. Throughout Chapters 5–7, though, I will discuss some of the specific software programs that can be used to conduct the various statistical tests.

t tests can actually be computed pretty easily by hand using a calculator unless the sample is exceedingly large. With a large sample size, the calculations may get cumbersome, with more room for errors. There are plenty of statistical software packages for calculating *t* tests. *t* tests can be calculated using Microsoft Excel – look under the Formulas tab and select "More Functions" ➜ "Statistical". If you have the Data Analysis Add-In, you can select the Data tab, then choose Data Analysis, and select from multiple variations of *t* tests in the drop-down menu.

In SAS, you can use the Proc T Test procedure or you can use the *t* tests task, which is an interactive, drop-down menu-style application in SAS Studio. You can also use Proc Univariate. In Table B.5, I provide a sample, fictional study analyzed using a single sample *t* test run with SAS.

In SPSS®, you would select the Analyze tab, choose "Compare Means" from the drop-down menu, and then select either Independent Samples *t* test or Paired-Samples *t* test, whichever is appropriate for your study. In R, you would use the t.test function. With Stata®, you can use the ttest command.

5.3 The NOVAs and Mixed Linear Model Analysis

First, a shout out to one of my students (Saba Masood) at UT Southwestern and her creativity in naming the various tests of ANOVA, "the NOVA's." The NOVA's are a set of statistical tests that use variance to assess differences between groups. In reality, even though variances are employed in the calculations of the NOVA's, the NOVA's really tell us about differences in the average, or mean, scores of groups. The outcome result of NOVA's is the *F* statistic, which is a ratio of the variance between groups divided by variance within the groups, also referred to as error variance. If there is more difference across the groups, variance across means, than there is variance within the groups, our *F* will be larger than 1.0, 1.0 being the average value in the *F* distribution. The further an obtained *F* value is from 1.0, the more likely it is to indicate a significant difference between groups. Like critical values of *t*, critical values of *F* are smaller for larger sample sizes, and larger from smaller sample sizes. Again, that is because larger sample sizes approach the population, and any *F* value much bigger than 1.0 is significant. Put another way, if you were to calculate *F* scores over and over for groups of samples of a certain size, assuming those samples were from the same population with the same characteristics, you would obtain a distribution of *F* scores, and that distribution would have a mean of 1.0, and the distribution would be positively skewed, with fewer and fewer obtained F values as you move away from 1.0 on the right side of the distribution. Larger sized samples will produce fewer F ratios above 1.0 than smaller sized samples in the event that the groups do not differ.

5.3.1 ANOVA

Simple ANOVA can be used to compare two groups of scores, just like a t test. In fact, the two statistics are directly, mathematically related to one another: when comparing two groups, $F = t^2$. The advantage of ANOVA over a t test is that you can compare more than two groups at a time. For instance, if you have three treatment groups, cognitive behavioral therapy (CBT), Medication, and Waitlist Controls, you can compare them on the dependent variable, a measure of depression, following the intervention period using ANOVA. With ANOVA, you can first determine if there are overall group differences, meaning variance in mean depression across the three groups beyond what you expect based on the overall variation of the individual participants. By conducting an ANOVA comparing the three groups, you obtain an F ratio. If that F ratio is statistically significant, you conclude that your groups differ in depression following the intervention period. You can then follow up your ANOVA with either planned comparisons or post hoc comparisons to figure out which pairs of groups differed, or which group differed from the others.

Planned and post hoc comparisons are individual comparisons between pairs of groups or between sets of groups. In the example above, you could compare CBT versus Medication, CBT versus Waitlist Control, and/or Medication versus Waitlist Control. You could also just compare the CBT group to the other two groups. Planned comparisons are comparisons that you plan to make in advance and for which you have specific predictions. Post hoc comparisons include every possible comparison between pairs of groups. There are multiple methods for calculating and interpreting both planned and post hoc comparisons, and each has different characteristics with different advantages and disadvantages.

A commonly used comparison is the Bonferroni t test, also referred to as the Dunn test. This test can be used for either planned or post hoc comparisons. This is a simple t test comparing pairs of means, but the α-level used for significance is divided by the number of comparisons made – this is the Bonferroni correction. The reduction in the required α-level is made in order to counteract the increase in Type I errors that will occur with multiple comparisons. That is, with multiple comparisons, you increase the likelihood of making at least one (or more) Type I error since each comparison carries a risk of Type I error (I talk more about the risk of Type I errors with multiple comparisons at the end of this chapter). The Bonferroni correction is considered to be conservative, and with more comparisons, the correction becomes stricter, decreasing the likelihood of concluding a significant difference between a pair of means. The advantage of the Bonferroni correction is that it is easy to calculate and it makes intuitive sense; however, if you have a lot of comparisons to make, you will be better off using a less conservative method.

The Tukey HSD (which stands for honestly significant difference) is a commonly used method for making post hoc comparisons. The Tukey HSD is also related to the t test for independent samples, though the denominator is a factor of the variance of the means with a correction for the number of comparisons. Fortunately, it is fairly simple to request Tukey HSD results when using a statistical software package to conduct an ANOVA.

In Table 5.8, I outline a sample study in which ANOVA would be appropriate for analyzing the data, provided the assumptions of ANOVA are met (to be discussed a little later). This is a single factor study, with one independent variable that has four levels.

Table 5.8 Sample study using ANOVA.

Independent variable: Treatment
   Levels:
      1: New adjunct medication
      2: Cognitive behavioral therapy (CBT)
      3: New adjunct medication + CBT
      4: Care as usual

Dependent variable: Beck Depression Inventory-II Score

Sample: All participants are patients in an outpatient psychiatry clinic diagnosed with major depression:
   50 are treated with the addition of a new medication
   50 are treated with CBT
   50 are treated with the addition of a new medication plus CBT
   50 are treated with their care as usual

Each patient is randomly assigned to either the control (care as usual) group, the new adjunct medication group, the CBT group, or the combined medication plus CBT group after consenting to be in the study. After a six-month treatment period, patients complete the Beck Depression Inventory-II.

An analysis of variance (ANOVA) is conducted to compare Beck Depression Inventory-II scores across the four groups to see if the four groups differ in the degree of self-reported depression following treatment. If the ANOVA is significant, the four groups can be compared to one another using either Bonferroni t tests or the Tukey's HSD test. This will help to determine which of the treatments was most effective.

5.3.1.1 ANOVA with a Multifactorial Design

In Section 5.3.1, I discussed using ANOVA to analyze group differences with one factor, or independent variable. In the example, I had one independent variable with four levels – the independent variable was the intervention type, and the four levels were CBT, Medication, CBT plus Medication, and Care as Usual. ANOVAs can also be used to analyze studies with a multifactorial design, that is, with more than one independent variable. In this case, you will obtain multiple F ratios, one for each independent variable, or main effect, as well as F ratios for each interaction (discussed further on Section 5.3.1.2).

So, back to the example above, let's say I am interested in the effect of intervention and whether there are differences between individuals with recurrent major depression and those with a first episode of major depression. I now have two independent variables, intervention type and chronicity of depression. This time around (for the sake of simplicity), let's say I have three levels of intervention type (CBT, Medication, and Care as Usual), while chronicity has two levels, Recurrent and Single Episode. This design is characterized as a 3×2 factorial between-subjects design. My ANOVA will produce three F ratios, an F for the main effect of intervention type, an F for chronicity, and an F for the interaction between intervention type and chronicity.

Before I explain what I mean by main effect and interaction, I want to note that ANOVAs can be conducted by hand, as in the olden days, or with readily available statistical software packages. A simple, one-way (single factor) ANOVA comparing two or more groups is actually quite easy to calculate by hand. A multifactorial ANOVA is not that much more difficult to calculate by hand, but it can be time consuming and cumbersome with lots of opportunity for making calculation errors.

5.3.1.2 Main Effects and Interactions

A main effect is an effect of an independent variable standing alone. So, in the example above, if the F for the main effect of intervention type is statistically significant, but the F for the main effect of chronicity is not significant, then we report that there were differences in depression scores across intervention type, but there was not a significant difference between the Recurrent and the Single Episode groups in their depression scores at the end of treatment.

But, do we know whether there was a differential effect of intervention for each of the chronicity types? For instance, did one type of intervention work better for one group (Recurrent) than the other (Single Episode) or vice versa? Whether there is a differential effect of one independent variable across levels of the other independent variable(s) is assessed by looking at the interaction effect. A significant interaction will tell us there are different effects of one independent variable at different levels of another independent variable, though it does not tell us how or in what direction the effect occurs at each level. To determine how the two (or more) variables interact, we will need to examine the mean scores of each 3×2 group. We can also analyze the difference between particular subgroups using planned or post hoc comparisons.

How many interaction effects are possible and to what degree? With two independent variables, you can have one interaction effect, the $A \times B$ interaction. With three independent variables, you will have three two-way interactions ($A \times B$, $B \times C$, and $A \times C$) as well as one three-way interaction ($A \times B \times C$). Sometimes, the three-way interaction is interpretable and meaningful, but as the layers increase, the interpretation becomes more complicated. More importantly, the sample size required to have adequate power to detect significant three-way (or higher) interactions becomes quite large.

Before we move on to the assumptions of ANOVA, let's go back to our example and discuss possible outcomes with regard to the interaction effect. A significant interaction in the face of a significant main effect of intervention but no overall main effect of chronicity can occur, for example, if one of the chronicity groups benefits from one intervention, while the other chronicity group benefits from another of the interventions, but not vice versa (see Figure 5.9). When reviewing the mean scores, we see that the interaction effect gives us important information that may affect how we treat patients. While, overall, participants showed more improvement following intervention in comparison to the care as usual group, which type of treatment was effective depended on their chronicity. The participants with recurrent major depression did not benefit if they participated in CBT, but they benefited if they

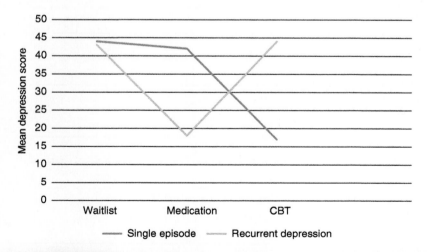

Depression score means	Waitlist	Medication	CBT
Single episode	44	42	17
Recurrent	43	18	44

Figure 5.9 Interaction between intervention and chronicity.

received medication. In contrast, the participants experiencing a single episode of depression benefited if they participated in CBT, but not if they were treated with medication. If we had not looked at the interaction effect, we would have come to the wrong conclusion about our interventions and our chronicity groups.

5.3.1.3 More Illustrations of Interactions and Main Effects

In Figure 5.10, I illustrate significant main effects of each treatment modality and chronicity in the absence of a significant interaction effect. In other words, in this study, the made-up data indicate that there were significant differences between the treatment groups in their depression scores. There were also significant differences between the single episode and recurrent depression groups. However, there was no interaction between the two independent variables. In other words, the effect of treatment modality was about the same for the two chronicity groups.

In Figure 5.11, I show you what an interaction looks like between two independent variables with two levels each. Here, we see that left-handed females receive higher scores on a handwriting quality scale administered by raters blinded to hand preference and gender than right-handed females, but left-handed males receive lower scores than right-handed males. Overall, there is not a significant main effect of either gender or hand preference.

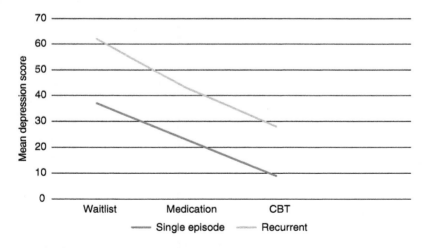

Depression score means	Waitlist	Medication	CBT
Single episode	37	23	9
Recurrent	62	43	28

Figure 5.10 Significant main effects of both intervention type and chronicity in the absence of an interaction.

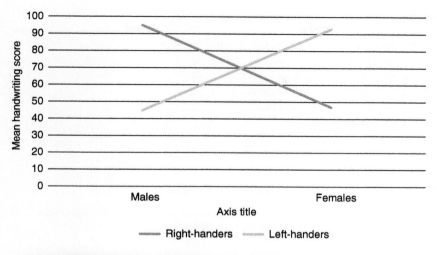

Figure 5.11 Significant interaction effect in the absence of main effects of gender or hand preference.

Handwriting score means	Right-handers	Left-handers
Males	95	45
Females	47	93

In Table 5.9, I summarize a sample study similar to the one discussed on the preceding pages. This can be characterized as a multifactorial study, in which there are two independent, between-subjects variables.

5.3.1.4 Assumptions of ANOVA
The main assumptions of ANOVA are very similar to the assumptions of t tests. They include:

1. Interval or ratio data
2. Independence of samples (the two or more groups are not related)
3. Normality (the distributions of the scores in each group are normally distributed)
4. Homogeneity of variance (the variance, or spread, of the scores in each group is approximately equal).

As I noted earlier when I introduced the concept of assumptions, some assumptions are stricter than others, depending on the statistical procedure. In general, ANOVA is fairly robust to violations of the third and fourth assumptions (normality and homogeneity of variance), though violations do diminish power. Violation of the assumption of normality is less impactful with larger samples, and violation of the homogeneity of variance assumption

Table 5.9 Example of a multifactorial ANOVA.

Independent variables: Treatment and chronicity
    Treatment levels:
        1: New adjunct medication
        2: Cognitive behavioral therapy (CBT)
        3: New adjunct medication + CBT
        4: Care as usual
    Chronicity levels:
        1: Single episode of major depression
        2: Recurrent major depression

Dependent variable: Beck Depression Inventory-II Score

Sample: All participants are patients in an outpatient psychiatry clinic diagnosed with major depression, half with their first, single episode, and half with chronic depression:
    25 participants with a single episode are treated with the addition of a new medication
    25 participants with chronic disease are treated with the addition of a new medication
    25 participants with a single episode are treated with CBT
    25 participants with chronic disease are treated with CBT
    25 participants with a single episode are treated with the addition of a new medication plus CBT
    25 participants with a chronic disease are treated with the addition of a new medication plus CBT
    25 participants with a single episode are treated with their care as usual
    25 participants with chronic disease are treated with their care as usual

Each patient from each level of chronicity is randomly assigned to either the control (care as usual) group, the new adjunct medication group, the CBT group, or the combined medication plus CBT group after consenting to be in the study. After a six-month treatment period, patients complete the Beck Depression Inventory-II.

A 4×2 analysis of variance (ANOVA) is conducted to compare Beck Depression Inventory-II scores across the groups to see if the treatment groups differ in their degree of depression (F test for the main effect of Treatment), and to determine whether the response to treatment modality differs depending on chronicity. The latter will be evaluated with the Treatment × Chronicity interaction F. If the ANOVA is significant, the individual groups can be compared to one another using one of the post hoc tests. If we are only interested in comparing each of the three treatment groups to care as usual, and without respect to chronicity, we would probably be OK using the more conservative Bonferroni t tests given the relatively small number of comparisons.

---

is less impactful when there are approximately equal numbers of individuals (e.g. participants or subjects) in each group or subgroup.

With reduction in power due to more severe violations of assumptions, especially when you have significant outliers in your groups, a nonparametric test may be better. These are reviewed in Chapter 6.

5.3.2 ANCOVA

Analysis of Covariance (ANCOVA) is a special case of ANOVA in which we try to statistically control for the effect of an extraneous variable. Extraneous variables add noise to our data, and using ANCOVA allows us to exert some control over that noise. ANCOVA accomplishes this by partialling out the impact the extraneous variable(s) has on the dependent variable. When we partial out an extraneous variable, we call that variable a covariate. A covariate is an extraneous variable that can be quantified, typically as a value on an interval or ratio scale. To be viable and useful, a covariate has to be a variable that correlates with, thereby directly impacts, the dependent variable. Another way to say that is that the covariate and the dependent variable have shared variance.

By partialling out the effect of the extraneous variable, we essentially remove that portion of the error variance (the noise) from our data. This means we are reducing the error term, the denominator, in the F statistic. This has the effect of increasing our F value, which is another way to say we have improved our power; we have increased our ability to detect group differences attributed to the independent variable of interest. This concept was illustrated earlier in Table 4.2.

There are times when partialling out the extraneous variable, the covariate, does not improve power, and may even reduce it. This occurs when the extraneous variable does not actually impact the dependent variable. In other words, if the extraneous variable is not correlated with the dependent variable, we will not be partialling out any noise. That is because the covariate's effect is normally partialled out from the shared variance between the covariate and the dependent variable; so, if there is no shared variance, there is no variance partialled out. At the same time, we will be reducing the size of our F statistic because we have used up an extra degree of freedom by including the covariate in our calculations, our model. The loss of the degree of freedom also changes the calculation of our F statistic, and in a detrimental way.

To illustrate what I mean, unfortunately I will have to bog you down just a bit with one more formula. That is the formula for F. F is a ratio, as you already know. It is the between-groups mean square (BG_{MS}) divided by the within groups mean square (the error term; WG_{MS}). The WG_{MS}, the error term in the F ratio, is basically the error variance divided by the within-groups degrees of freedom. The within-group degrees of freedom is the total number of participants (N) minus 1, minus the number of groups, and minus 1 for each covariate. So, when we use a covariate, we reduce the degrees of freedom used in calculating WG_{MS}. This will result in an *increase* in the size of WG_{MS} because the divisor in calculating it is reduced. Since the F statistic is a ratio of the BG_{MS}/WG_{MS}, an increase in the size of WG_{MS} reduces the size of our F statistic, and we will be less likely to be able to reject our null hypothesis.

You don't need to bog yourself down with the specifics of the calculations – unless you really want to. For now, the point is that using a covariate can reduce power if the covariate does not actually impact the dependent variable, that is, if the covariate does not correlate with the dependent variable. That is because we have done nothing to reduce our error term in the F ratio, the denominator, and instead have increased the error term by using up a degree of freedom for that wasted covariate.

Before I list the assumptions of ANCOVA, let's consider an example in which ANCOVA would be useful. Let's say we want to compare mean starting salary of graduates with a Liberal Arts degree versus graduates with a Business degree. If the purpose of the research is to determine whether choice of major impacts starting salary, then we will want to control for extraneous variables that could also relate to mean starting salary. If there is a known or suspected linear relationship (i.e. a linear correlation) between SAT® scores and mean starting salary (but not between SAT scores and choice of major), then it would be wise to control for the effect of SAT scores, which could confound any conclusions we have about choice of major and starting salary. In this case, we could conduct an ANCOVA, treating SAT score as our covariate. We could first run a correlation analysis between SAT and salary to make sure we are not reducing power by including SAT in the model unnecessarily. If the correlation is significant, then we are probably safe running the ANCOVA. In fact, we may be wise to run an ANCOVA so that we can control for this extraneous variable, SAT, which could otherwise confound our results.

The assumptions for ANCOVA are the same as for ANOVA with a few additions. These major additional assumptions are:

1. A linear (i.e. straight positive or negative) relationship between the covariate and the dependent variable.
2. Homogeneity of regression slopes (the relationship between the covariate and the dependent variable is the same across groups).
3. Independence of the covariate and the independent variable (they don't correlate).

Good characteristics of covariates and their use:

1. They are actually correlated with the dependent variable, and the correlation is linear.
2. They are reliably measured and valid.

Table 5.10 Example of a single-Factor ANCOVA.

Independent variable: Treatment

Levels:

  1: New adjunct medication

  2: Cognitive behavioral therapy (CBT)

  3: New adjunct medication + CBT

  4: Care as usual

Covariate: Age

Dependent variable: Beck Depression Inventory-II Score

Sample: All participants are patients in an outpatient psychiatry clinic diagnosed with major depression:

  50 are treated with the addition of a new medication

  50 are treated with CBT

  50 are treated with the addition of a new medication plus CBT

  50 are treated with their care as usual

Each patient is randomly assigned to either the control (care as usual) group, the new adjunct medication group, the CBT group, or the combined medication plus CBT group after consenting to be in the study. After a six-month treatment period, patients complete the Beck Depression Inventory-II.

An analysis of covariance (ANCOVA) is conducted to compare Beck Depression Inventory-II scores across the four groups while controlling statistically for the effects of age. The analysis is conducted to see if the four groups differ in the degree of self-reported depression following treatment. If the ANCOVA is significant, the four groups can be compared to one another using either Bonferroni *t* tests or the Tukey's HSD test.

---

3. A minimal number of covariates is included.
4. There is minimal or no correlation among covariates when more than one is included.

In Table 5.10, I describe a variation of the example in Table 5.8. In this study, I have added a covariate, age, because it turns out that the groups were not identical in their average age, and age correlated with Beck Depression Inventory-II scores in this group of patients.

5.3.3 MANOVA/MANCOVA

Multivariate Analysis of Variance (MANOVA) is another special case of ANOVA. Here, we can assess more than one, related dependent variable at a time. We might want to conduct a MANOVA instead of moving straight to separate ANOVAs for each dependent variable we are evaluating, particularly when our dependent variables measure a similar construct. In doing this, we can reduce our risk of making Type I errors that would be caused by making multiple comparisons. Recall that performing multiple comparisons increases our risk of obtaining a significant result at a given p value when there is no true effect. If the MANOVA is significant, we can explore our data with additional analyses to determine which dependent variables were most influenced by the independent variable(s). However, when conducting these post hoc tests, we still have to be mindful of the elevated risk of Type I errors by adjusting our p values.

There are different options for exploring our data further if we obtain a significant result with the initial MANOVA. Three of these are: follow-up ANOVAs (also called univariate ANOVA), Discriminant Function Analysis, and the Roy–Bargmann Stepdown Procedure. Follow-up ANOVAs are probably the most intuitive option, and this option is often found in the literature. In this case, each dependent variable is analyzed with a separate ANOVA. The key is to manage the elevated risk of Type I errors through a correction, such as the Bonferroni correction.

While follow-up ANOVAs are commonly performed, purists seem to prefer the other options, especially the Discriminant Function Analysis. Just briefly, the Discriminant Function Analysis informs the researcher about which of the dependent variables is most strongly related to, or impacted by, the independent variable. The Roy–Bargmann Stepdown Procedure is similar to the simple univariate ANOVAs, except the researcher selects the order in which each univariate dependent variable is evaluated a priori. After the first dependent variable is evaluated with an ANOVA, it becomes a covariate in the evaluation of the next dependent variable. This step is repeated for each successive dependent variable in the list as long as each ANCOVA is significant. This procedure is described in detail by Holmes Finch (2007). The idea is that the effect of each dependent variable is partialled out in each subsequent step as a means to address the relationship between those related dependent variables.

As with ANOVA, a researcher can control for extraneous variables when using MANOVA by treating the extraneous variable as a covariate. This is classified as multivariate analysis of covariance (MANCOVA). In this case, there would be at least one independent variable, more than one dependent variable, and at least one covariate.

The assumptions of MANOVA and MANCOVA are similar to the assumptions of ANOVA and ANCOVA, respectively, with these additions:

1. Multivariate normality – all linear combinations of the dependent variables are normally distributed; loosely, this means for each given value of one dependent variable, another dependent variable will be normally distributed. MANOVA is fairly robust to deviations from this assumption as long as there are no substantial outliers.

2. Homogeneity of covariance matrices – also loosely, the correlations between the dependent variables are about the same for each group. MANOVA is fairly robust to deviations from this assumption when using fairly equal sample sizes.
3. Linearity of the relationships between the dependent variables.
4. Absence of multicollinearity (excessive correlation, about greater than 0.8 or 0.9 between the dependent variables), though they should be moderately correlated.

In Table 5.11, I offer an example of a study in which MANOVA is used to analyze the data. This is a single-factor study with four levels of the independent variable, and three, related dependent variables.

Table 5.11 Example of a multivariate analysis of variance study.

Independent variable: Treatment

Levels:

1: New adjunct medication

2: Cognitive behavioral therapy (CBT)

3: New adjunct medication + CBT

4: Care as usual

Dependent variables: Beck Depression Inventory-II Score, Beck Anxiety Inventory (Beck, Epstein, Brown, & Steer, 1988) Score, and the Hamilton Depression Rating Scale (Hamilton, 1960)

Sample: All participants are patients in an outpatient psychiatry clinic diagnosed with major depression:

50 are treated with the addition of a new medication

50 are treated with CBT

50 are treated with the addition of a new medication plus CBT

50 are treated with their care as usual

Each patient is randomly assigned to either the control (care as usual) group, the new adjunct medication group, the CBT group, or the combined medication plus CBT group after consenting to be in the study. After a six-month treatment period, patients complete the Beck Depression Inventory-II and the Beck Anxiety Inventory, which are both self-report measures. A clinician blind to each patient's treatment group assignment completes the Hamilton Depression Rating Scale, which is an observer report.

A multivariate analysis of variance (MANOVA) is conducted to compare the groups on their overall degree of depression and anxiety as measured by the three test instruments to see if the four groups differ following treatment. If the MANOVA is significant, we plan to specifically look at differences between each of the three treatment groups and the control group on self-reported depression (BDI-II scores) and on Hamilton Depression Rating scores using Bonferroni *t* tests keeping in mind that the correction will be somewhat conservative given the number of comparisons (which will total six).

5.3.4   Statistical Software Packages for Conducting ANOVA/ ANCOVA/MANOVA

If you are really ambitious, love detailed math calculations, or are stranded on a desert island without access to a computer, you could, technically, conduct an ANOVA or its relatives by hand. However, if none of these scenarios describes you, have no fear. Statistics packages are here.

If you prefer SAS for your analyses, you could use Proc ANOVA for some of the NOVAs, though Proc GLM (General Linear Models) offers more flexibility and options. You could also use a drop-down menu-style application in Tasks with SAS Studio – there is One-Way ANOVA, N-Way ANOVA, and Generalized Linear Models. In Appendix B of this primer, in Tables B.1–B.4, I review the steps used to conduct an ANOVA in SAS using Proc GLM.

With SPSS, you could again select the Analyze tab, choose Compare Means, and then select One-Way ANOVA. For the more complex analyses, you would select the Analyze tab and then choose General Linear Models. You would then choose between Univariate, Multivariate, or Repeated Measures. In R, you could use the aov() function. For more complex analyses, you would use the glm() function. In Stata, using the graphical user interface (GUI), you first select the Statistics tab, then select "Linear models and related" ➔ "ANOVA/MANOVA" and then within the next drop-down menu, you select the appropriate form of ANOVA. Finally, I should note that if you have the Data Analysis Add-In for Excel, you can do some basic ANOVAs, though you will find a lot more flexibility and options with one of the above statistical software packages.

5.3.5   Repeated Measures: ANOVA-RM and Mixed Linear Model Analysis

In Chapter 3, I reviewed within-subjects research designs, also referred to as repeated measures. Earlier in this chapter, I discussed how a matched pairs *t* test can be used to analyze a study with repeated measures at two points. When we repeat a measure two *or* more than two times per participant (e.g. baseline, post intervention, and later follow-up), or when we also have a between-subjects variable, we can instead use ANOVA-Repeated Measures (ANOVA-RM) or Mixed Linear Model Analysis to analyze our repeated measures data. Some researchers even use ANCOVA when they just have pre- and post-testing as their repeated measure, and they treat the pretest measure as a covariate. These three methods are briefly reviewed below.

5.3.5.1   ANOVA-RM
ANOVA for analyzing repeated measures (ANOVA-RM) is relatively easy to conduct and intuitive. Most statistical software packages handle ANOVA-RM. The repeated measure is listed as an independent variable, but with an

indicator that the measure is repeated. In this way, the calculations are able to factor in the relationship, or correlation, between the levels of that independent variable. The drawback of ANOVA-RM is that any missing data for a participant will result in that participant being dropped from the analyses. This could bias the results and will also decrease the sample size and, thus, the power. Another drawback of ANOVA-RM is its susceptibility to violations of assumptions, particularly the assumption of sphericity (discussed at the end of this subsection). And, when participants are dropped from the analyses due to missing data, the adverse impact of deviations from the assumptions increases given the diminishing sample size.

In terms of research design, within-subjects studies allow us to manage some of the noise caused by extraneous variables by using participants as their own controls. Statistically, the advantage of the repeated measures ANOVA (versus a between-subjects design) is that we remove some of the within groups variance from the error term in calculating F. This, as has been noted, will increase our power by increasing the F ratio.

When we have a between–within subjects design, such as when we are comparing treatment modality (e.g. drug versus placebo) across time (baseline, post-treatment, and follow-up), the primary effect of interest will be the interaction between treatment modality group and time (this is illustrated in Figure 5.12). That is because we are interested in learning whether there is a difference between the treatment groups in their change from baseline to Time 2 and Time

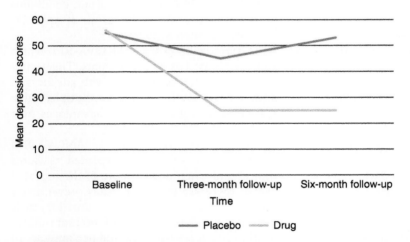

Depression score	Baseline	Three-month follow-up	Six-month follow-up
Placebo group	55	45	53
Medication group	56	25	25

Figure 5.12 Significant time × treatment group interaction in a repeated measures study.

3. In the study illustrated in Figure 5.12, we see that the drug under study was associated with more improvement from baseline to the three-month follow-up appointment than was the placebo, and this improvement maintained through to the six-month follow-up appointment.

The assumptions of ANOVA-RM are very similar to those of ANOVA. One assumption unique to ANOVA-RM is sphericity. In a nutshell, the assumption of sphericity requires that the variances of the difference scores between pairs of any and all combinations of the repeated measure are approximately the same. When running an ANOVA-RM using a statistic package, the assumption of sphericity can be formally tested to some degree by requesting Maulchy's test, and there are options for correcting for violations of the assumption. Maulchy's test may under-report violations of sphericity for smaller samples, though smaller samples with few repeated measures can be tested by simply calculating the variances of the difference scores and comparing those directly.

5.3.5.2 Mixed Linear Model Analysis

Mixed Linear Model Analysis can also be used for analyzing repeated measures data, and it has some definite advantages over ANOVA-RM. It is very similar to ANOVA-RM, including that it results in F statistics, and the Group × Time interaction tells the story of whether there is a different trajectory over time between our groups; that is, whether our experimental group changes more than, say, our control group. In fact, under certain conditions (e.g. with a basic design and absence of missing data), the results of ANOVA-RM and Mixed Linear Model Analysis will often be identical.

Mixed Linear Model Analysis is more complex than ANOVA-RM, which explains why many of us did not learn about it as young students. The usefulness of Mixed Linear Model Analysis has increased greatly with advances in computer technology and statistical software packages. It is very flexible, which contributes to its complexity; however, this gives it many advantages over ANOVA-RM.

Importantly, Mixed Linear Model Analysis is superior to ANOVA-RM when there are missing data. Missing data may occur in a repeated measures study if a participant is missed or absent when one or more of the repeated measurements is obtained or when there are technical issues preventing a measurement. When participants are absent or when there is attrition, their data are no less important. As was noted in the discussion of internal validity, attrition can bias results particularly when it is systematic and not random. In ANOVA-RM, when a participant or subject has missing data, they fall victim to listwise deletion; that is, the entirety of that individual's data are dropped from the analysis if they have even one piece of missing data from the repeated measure. Mixed Linear Model Analysis handles missing data without listwise deletion; participants are not dropped from the analysis due to missing data.

Another rather cool advantage of the Mixed Linear Model Analysis is the ability to treat time as a continuous variable. This allows us to analyze the trend over time, by calculating the correlation between time and the dependent variable. In this case, differences in spacing over time are also managed so that they are not equated; in other words, when the time points are not equally spaced, that is accounted for and that spacing is not treated as though it is the same across the repeated measure.

Mixed Linear Model Analysis is also better suited when the repeated measures are not balanced or when you want post hoc analyses. A good explanation of the advantages of Mixed Linear Model Analysis can be found online with the Analysis Factor (The Analysis Factor, 2008–2018a).

5.3.5.3   ANCOVA

Just a quick note about ANCOVA in the analysis of repeated measures: some researchers will use ANCOVA for repeated measures when the dependent variable is only measured at two time points – baseline and post intervention, for example. The baseline measure is treated as a covariate. While mathematically this makes some sense, the results are not directly answering the question at hand in most cases. Typically, when we have repeated measures, we want to know if there is change over time. ANCOVA does not answer that question. ANCOVA tells you whether groups (the between-subjects variable) differ at the end of the study period while controlling for any pre-existing differences (their baseline, pretest measure). It does not tell you if the groups changed differentially, or in what direction that change occurred for each group. Interestingly, when ANCOVA is used for repeated measures in this way, researchers often follow up with a planned comparison or post hoc t-test to compare scores at baseline and post-testing within each group. This same information can be obtained in one step with ANOVA-RM or Mixed Linear Model Analysis with the Group × Time interaction without running the risk of elevating the Type I error.

In Table 5.12, I illustrate an example of a repeated measures ANOVA with two between-subjects variables and one within-subjects (repeated) measure. This is a $4 \times 2 \times 3$ Repeated Measures design. This particular example is a variation of the study reported in Table 5.9.

5.3.5.4   Statistical Software Packages for Conducting Repeated Measures Analyses

Even if you love doing math, I don't think I could in good conscience recommend doing a repeated measures analysis by hand, especially a Mixed Linear Model Analysis. If you like using SAS, you can turn to Proc GLM again, this time adding a "Repeated" statement. For the more flexible and sophisticated Mixed Linear Model Analysis, you can use Proc Mixed procedure, again using a "Repeated" statement. You could also use Mixed

Table 5.12 Example of a repeated measures study.

Independent variables (Between): Treatment and chronicity

Independent variable (Within): Time

Treatment levels:

   1: New adjunct medication

   2: Cognitive behavioral therapy (CBT)

   3: New adjunct medication + CBT

   4: Care as usual

Chronicity levels:

   1: Single episode of major depression

   2: Recurrent major depression

Time levels:

   1: Baseline

   2: Immediate post-treatment

   3: Six-month follow-up

Dependent variable: Beck Depression Inventory-II Score

Sample: All participants are patients in an outpatient psychiatry clinic diagnosed with major depression, half with their first, single episode, and half with chronic depression:

   25 participants with a single episode are treated with the addition of a new medication

   25 participants with chronic disease are treated with the addition of a new medication

   25 participants with a single episode are treated with CBT

   25 participants with chronic disease are treated with CBT

   25 participants with a single episode are treated with the addition of a new medication plus CBT

   25 participants with a chronic disease are treated with the addition of a new medication plus CBT

   25 participants with a single episode are treated with their care as usual

   25 participants with chronic disease are treated with their care as usual

Each patient from each level of chronicity is randomly assigned to either the control (care as usual) group, the new adjunct medication group, the CBT group, or the combined medication plus CBT group after consenting to be in the study. At baseline, again following the six-month treatment period, and again at a six-month follow-up, patients complete the Beck Depression Inventory-II.

This is a $4 \times 2 \times 3$ mixed factorial design, and can be analyzed either with repeated measures analysis of variance (ANOVA-RM), since we are lucky enough not to have any missing data and we meet all the required assumptions, or with Mixed Linear Model analysis. With these analyses, we will be able to compare Beck Depression Inventory-II scores across the groups and across time. The purpose is to see if the treatment groups differ in their change in depression scores over time, including whether that trend maintains at follow-up. The Treatment × Time interaction will be the most critical analysis in answering that question, though we are also interested in knowing whether the relative improvement depending on treatment modality is the same for the patients with chronic versus single-episode depression.

Models found in the Tasks menu of SAS Studio for a drop-down menu application. In that application, the repeated variable will be specified in the Model tab under Model Effects.

To perform an ANOVA-RM with SPSS, you again select your trusty "Analyze" tab, then choose "General Linear Model" which will open another drop-down menu from which you will pick, "Repeated Measures." In SPSS, to perform the Mixed Linear Models version of a repeated measures analysis, you will again select the Analyze tab, but then choose "Mixed Models" from the drop-down menu followed by "Linear." That will bring up a dialog box ("Linear Mixed Models: Specify Subjects and Repeated") where, among other items, you will be able to specify your repeated variable.

If you are using R and it is most appropriate to analyze your repeated measures data with a Mixed Linear Models analysis, you can use the lmer function available in the lme4 package. If ANOVA-RM works with your variables and data, you can use the aov function or ezANOVA functions, depending on various parameters (e.g. the ezANOVA is for within-subjects designs).

In Stata, for an ANOVA-RM, you would again go to the Statistics tab, select "Linear models and related" followed by "ANOVA/MANOVA" → "Analysis of variance/covariance." Then, in the ensuing dialog box, under the "Model" tab, you can specify your repeated variable. For a Mixed Linear Model Analysis, you would probably want to use the "xtmixed" command.

One thing to be aware of is that most statistical software packages are looking for data to be stored in long form for Mixed Linear Models analysis of repeated measures. This means, in a nutshell, that instead of recording all the repeated measures for a participant in the same row, a different row, or record, is stored in the data set for each repeat of the repeated variable – so, each participant would have one row for the baseline measure, for instance, one row for the 6-month measure, and one row for the 12-month follow-up measure. Don't worry if you have your data stored in wide form (e.g. one row per participant with baseline, 6-month, and 12-month follow-up measurements all together in the same row) because most statistical software packages have a series of commands or code that allow you to transform your data set from the long format to the wide format.

5.3.6 Summary

That wraps up our discussion of the flexible and powerful NOVAs. Many of the research designs discussed in Chapter 3 lend themselves to analysis using the NOVAs. However, there are many other options when planning studies and analyzing data. Many of these are conceptualized in terms of the *relationships* between variables as opposed to the *differences* between them. These types of analyses are reviewed next.

5.4 Correlation and Regression

So, now we will take a sharp turn in another direction. Or will we? Correlation and regression allow us to analyze our data differently than ANOVA and *t* tests, but they are really not much more than a different way to talk about data. Mathematically, the NOVAs and regression are related. I personally prefer ANOVA because it makes more intuitive sense to me, but other people are more comfortable thinking of their data in terms of relationships and prediction. Correlation tells us if and how variables relate, or covary with one another, and regression allows us to determine whether and how strongly a variable or set of variables predicts scores on another variable or set of variables. While I prefer the NOVAs in most scenarios, there are also times when a regression technique has advantages over its counterpart in the NOVAs.

5.4.1 Correlation and Multiple Correlation

Correlation is a measure of the relationship between two variables, and multiple correlation is a measure of the relationship between one variable and a set of additional variables. The calculation of correlation is based on the degree to which two variables covary; that is, as one variable for one participant or observation varies, the other variable changes to a similar degree, either in the same or the opposite direction (as in a negative correlation). A positive correlation between two variables indicates that, as one variable increases in value, the other tends to increase as well. A negative correlation between two variables indicates that, as one variable increases in value, the other tends to decrease. The strength of this relationship, either positive or negative, is indicated by the size of the correlation coefficient, which can range from −1 to +1, with 0 indicating no linear relationship and +1 or −1 indicating a perfect, consistent relationship across observations (participants). A simple correlation between two continuous variables (e.g. the relationship between height and weight) is represented by Pearson's r. We can also calculate a point–biserial correlation between one continuous variable and one binary or dichotomous variable (e.g. the relationship between height and gender).

A multiple correlation between one variable and a set of additional variables is represented by R, which can range from 0 to +1. For instance, we might be interested in the relationship between endurance and two additional variables, such as Body Mass Index (BMI) and hours exercised per week. When we have multiple predictor variables (such as BMI and exercise), we can also calculate partial correlations to consider each variable's relationship with the criterion variable with the effect of the other(s) held constant for both the criterion and predictor of interest. This basically allows you to see the relationship between two variables (e.g. endurance and exercise) while controlling for the effects of a third, potentially confounding variable (e.g. BMI), that could be contributing to the correlation between the first two. Semi-partial correlation is similar, but the effect of the extra predictor is only held constant for the criterion variable.

The assumptions of the Pearson correlation are:

1. Random and independent selection with pairs of data for each observation (i.e. each observation must include a value for each of the two variables).
2. Interval or ratio level of measurement for both variables.
3. Linear relationship between the variables (not curvilinear).
4. Normal distribution of each variable with no significant outliers.
5. Homoscedasticity (the variance of the first variable is equal at each value of the second variable, and vice versa).

In addition to being aware of the assumptions involved in correlations, it is important to be aware that restricted ranges in scores will reduce the strength of a correlation. That is, if your range of scores is restricted, or truncated, such that you are only considering a portion of scores (i.e. only high values, only low values, or only midrange values), your correlation coefficient will be small, regardless of whether there is a true relationship between your two variables in the population. This is a common phenomenon, for example, when studies address the relationship between entrance exam scores and later grade point average (GPA) when only those with higher entrance scores are admitted into the academic program being considered. If all entrance examinees were admitted into a program, the true correlation between the entrance exam score and GPA would likely be higher than a correlation using only the admitted students with the higher range of entrance exam scores. Since correlation is a measure of covariance between pairs of scores, any reduction in variance (such as due to restriction of range) will reduce the covariance.

In Tables 5.13 and 5.14, I present examples of studies with correlational analyses. The first is a simple correlation between two variables, age and wisdom, and the second is a multiple correlation between depression scores and three additional variables.

5.4.2 Regression and Multiple Regression

Regression, as noted, allows us to determine if scores on a variable (the predictor or independent variable) predict scores on another variable (the criterion or dependent variable). In multiple regression, we can determine to what degree a set of predictor variables predicts a criterion variable. The technique also allows us to determine to what degree each predictor variable contributes to the overall prediction. Multiple correlation and multiple regression are directly related, and a multiple regression analysis will result in a multiple correlation coefficient, R. In addition, when we use a statistics program to run a multiple regression, we will be able to obtain an F statistic, which will tell us whether we have a significant model, that is, a significant set of predictor

Table 5.13 Sample study of a correlation between two variables.

Variables: Age and score on a test of wisdom
  Type and level of measurement:
  Age: Ratio
  Standardized Wisdom test score: Interval

Sample: 200 individuals are recruited from the general population in multiple cities and towns across the country. Recruitment occurs via Twitter solicitation, at malls, and through direct mailing. Individuals are excluded from the study if they have a history of brain injury, dementia, or stroke, or if they have been previously diagnosed with intellectual disability.

Each participant reports their age, verified with formal identification, and completes a test of wisdom designed and validated by the researchers.

A Pearson correlation is calculated between age and wisdom score to tell us if there is a significant relationship between age and wisdom. A significant, positive relationship would tell us that increasing age (i.e. the older the person) is associated with increasing wisdom. A negative correlation (boo!) would tell us that increasing age is associated with decreasing wisdom. As a reminder, a correlation does not prove causality, only an association, or relationship. There may be a third, unrelated variable contributing to the relationship.

variables for informing us about the value of the criterion variable. We can also obtain separate *t* tests that tell us whether the contribution of each individual predictor variable, or independent variable, is significant. We can additionally create an equation that can be used to predict scores on the criterion variable based on values of the independent variables (see Table 5.15). This equation includes multipliers for each independent variable; these are referred to as coefficients when the predictor and criterion variables are not standardized, and β weights when they are standardized variables. These coefficients, or weights, reflect the degree to which a predictor variable correlates with the criterion separate from the other predictor variables (i.e. with their influence partialled out).

There are three major types of multiple regression. These are: simultaneous multiple regression, hierarchical multiple regression, and statistical multiple regression, reflecting when and how each independent variable is introduced into the model and analyzed. Determining when each predictor variable is added to the model for analysis depends on pre-existing knowledge and theory.

A nice component of regression is its usefulness in assessing the effects of moderators, mediators, and confounding variables. A moderator is a variable that interacts with a predictor variable in influencing the criterion variable. Moderators can enhance, buffer, or reverse the effect of the predictor variable. A mediator explains the relationship between two other variables; it tells us whether the relationship between the predictor and criterion can be

Table 5.14 Sample study using multiple correlation.

Criterion variable: Beck Depression Inventory-II Score

Predictor variables: Total hours of cognitive behavioral therapy, typical minutes of exercise per week, and typical hours spent volunteering per week

    Type and level of measurement

    Beck Depression Inventory-II: Ratio

    Hours of cognitive behavioral therapy: Ratio

    Minutes of exercise per week: Ratio

    Hours spent volunteering per week: Ratio

Sample: 105 participants are recruited from an outpatient psychology/psychiatry clinic (I used G*Power to plan my sample size conservatively estimating a population correlation of about 0.1, and setting power to 0.8 and $\alpha = 0.05$). Hours of cognitive behavioral therapy in the clinic is obtained from the clinic records, and a research assistant interviews participants to obtain the self-reported minutes of exercise per week and hours spent volunteering. The Beck Depression Inventory-II is also administered.

All variables are normally distributed, and none are restricted in range (phew). A multiple correlation is calculated to determine whether and how strongly the three predictor variables are correlated with depression scores. We can also look at the correlation matrix (a table showing the correlations between each pair of variables) to determine the strength and direction of the correlations – we are likely predicting negative and significant correlations, with more exercise, more volunteer work, and more hours of cognitive behavioral therapy being associated with lower depression scores. But, we can also calculate partial correlations to consider the independent contribution of each variable to depression scores (since the three predictor variables may be correlated with one another). Even better, we could just perform a Multiple Regression analysis...to be continued...

Note: A strong correlation, remember, does not prove causality. Just because we label the depression scores as our criterion variable does not mean the predictor variables *caused* the depression severity. In fact, it could be that depressed people are less likely to exercise, work, or participate in therapy and not necessarily that those factors reduce depression. We would have to conduct an experiment with a control group to suggest the presence and direction of causality.

Table 5.15 Multiple regression equation.

$$\hat{Y} = b_0 + b_1 X_1 + b_2 X_2 + \cdots b_i X_i$$

where $\hat{Y}$ is the predicted value of the criterion variable, b_0 is the intercept (the constant), and the remaining b's are the regression coefficients for the respective independent variables, labeled X_1 through X_i.

---

explained as being due to a third variable. That third variable is the mediator; the predictor variable impacts the value of the mediator, which, in turn, impacts the value of the criterion. A confound in regression analysis is a variable that correlates with both the predictor and the criterion variable such that a spurious relationship between the predictor and criterion is created. In this case, there is no causality between the predictor and the criterion; rather, their relationship is coincidental. A discussion of the how-tos of conducting analyses of moderators, mediators, and confounds is beyond the scope of this primer, but there are many online articles and resources for conducting these analyses with SAS or SPSS.

The assumptions for multiple regression are similar to assumptions we have discussed for some of the NOVAs. They include:

1. Interval or ratio data for the criterion variable
2. Independence of observations (participants)
3. Linearity (versus curvilinearity) in the relationships between the variables.
4. Homoscedasticity – the variance of criterion scores around the regression line should remain fairly constant at all values of a particular predictor variable
5. Absence of multicollinearity (strong correlations between predictor variables) or singularity (perfect correlation between predictor variables)
6. Absence of outliers or other highly unusual points
7. Normality of the residuals (i.e. the errors or distance of each criterion variable from the predicted criterion given the predictor variables).

And, finally, as with designing any study it is important to have an adequate sample size. Just as with ANOVA, determining sample size will be influenced by the number of predictor variables. There is really no prefect rule of thumb for determining the right number of observations (participants) as there are many moving parts, including the number of variables, desired power, anticipated effect size, and α. Fortunately, there are online calculators that can be used to determine the best sample size given these parameters.

In Table 5.16, I demonstrate a sample multiple regression study. This example is a natural extension of the study outlined in Table 5.14.

5.4.3 Statistical Software Packages for Conducting Correlation and Regression

Once again, for those math calculation enthusiasts with nimble calculator fingers, simple correlations can be calculated by hand. However, with a lot of data, there are a lot of calculations, squaring, adding, adding and squaring, etc. So, if you have access to a spreadsheet like Excel, or a statistical software package, I would recommend going one of those routes. In most cases, to calculate the correlation between two variables, you will use a correlation procedure, but when you want to calculate a multiple correlation, you will use a regression procedure.

Table 5.16 Sample multiple regression as an extension of the study in Table 5.14.

Criterion variable: Beck Depression Inventory-II Score

Predictor variables: Total hours of cognitive behavioral therapy, typical minutes of exercise per week, and typical hours spent volunteering per week

    Type and level of measurement

        Beck Depression Inventory-II: Ratio

        Hours of cognitive behavioral therapy: Ratio

        Minutes of exercise per week: Ratio

        Hours spent volunteering per week: Ratio

Sample: 100 participants are recruited from an outpatient psychology/psychiatry clinic. Hours of cognitive behavioral therapy in the clinic is obtained from the clinic records, and a research assistant interviews participants to obtain the self-reported minutes of exercise per week and hours spent volunteering. The Beck Depression Inventory-II is also administered.

All variables are normally distributed, and none are restricted in range, there are no outliers, and the predictor variables correlate only modestly with one another. A multiple regression analysis is performed to determine whether exercise, cognitive behavioral therapy, and volunteer time together predict the level of depression as measured by the Beck Depression Inventory-II. The significance of the "model" will be demonstrated through an F test generated by the analysis, and we will also be able to see the amount of variance in Beck scores that is accounted for by the other three variables (this will be depicted as R^2). We can also see the degree to which each of the predictor variables are uniquely related to the depression scores, and whether each relationship is significant. We will additionally be able to see whether the relationship between each predictor score and the depression scores is negative or positive. Negative partial correlations would indicate that more of each is associated with a lower depression score; alternatively, less of each (therapy, exercise, or volunteer work) is associated with higher depression scores.

In Excel, you can select the Formulas tab, choose "More Functions" → "Statistical" → "Correl." If you have the Data Analysis Add-In, you can select the Data tab, select Data Analysis, and then select Correlation from the drop-down menu (Regression is also available in the Add-In).

With SAS, you can use the Proc Corr procedure. If you want the multiple correlation, you will use Proc Reg, which actually performs Regression analyses. In Appendix B, I provide a sample, fictional multiple regression analysis run using SAS in Table B.7.

In SPSS, you will still select the Analyze tab, then choose "Correlate." You will have the option of selecting bivariate (a simple correlation). For multiple correlation (and, actually regression), choose "Regression" from the Analyze tab and then choose "Linear." In R, you can use the "cor" function for bivariate (two variable) correlations, and the "lm" function for multiple regression. With Stata using the GUI, you will select the Statistics tab, choose "Summaries, tables, and tests" from the drop-down menu ➜ "Summary and descriptive statistics" ➜ "pairwise correlations" or "Correlations and covariances." For a regression analysis, you again select the Statistics tab, but then choose "Linear models and related" ➜ "Linear regression."

5.5   Logistic Regression

Logistic regression is similar to multiple regression, except it is used to predict group membership in a category rather than predicting a value on a criterion variable measured on an interval or ratio scale. Binary logistic regression is used when the categorical variable has two levels (e.g. success versus failure, exerciser versus sedentary, or continue versus discontinue). In multinomial logistic regression, the categorical variable has more than two levels (e.g. diabetic, prediabetic, or nondiabetic), which are usually ordinal in nature. The predictor variables are variables that may influence whether a participant falls into one category or another (e.g. positive family history of diabetes, weekly calorie consumption, and hours of exercise per week). Logistic Regression Analysis allows us to determine how many and which of our predictor variables matter in influencing membership in a category as well as how strong that influence may be.

Ultimately, logistic regression provides estimates of the odds of an event as well as the adjusted odds ratio. The odds tells us the likelihood that an individual with a given set of predictor variable scores will have one outcome versus the other on the criterion variable. The adjusted odds ratio tells us to what extent an increase in the value of one specific predictor variable will increase the likelihood of one outcome or the other while holding the other predictor variables constant.

Logistic regression is somewhat akin to MANOVA in reverse, but it has some advantages in terms of the assumptions required. The assumptions in logistic regression are:

1. The criterion variable is categorical, not continuous
2. Independence of observations (participants)
3. Absence of specification errors (no important predictors are omitted, and no irrelevant predictors are included)
4. Absence of multicollinearity (no strong relationships between the predictor variables)
5. Linearity between the continuous predictor variables and the logit transformation of the criterion variable (in other words, the predictor variable(s) may need to be transformed in a way that results in a linear relationship between the transformed values and the already transformed criterion; the transformation you perform will be determined by the nature of the relationship between the variables).

As indicated above, logistic regression does not require some of the assumptions that are important in MANOVA. Those include multivariate normality and homogeneity of covariance matrices. A disadvantage of logistic regression, though, is the need for a large sample size.

In Table 5.17, I review an example of a study analyzed using Logistic Regression. In this made-up, but realistic study, the purpose is to determine what factors predict success in smoking cessation.

Table 5.17 Example of a Logistic Regression Analysis.

Criterion variable: Smoking cessation
  Levels:
    1: Successful cessation
    2: Unsuccessful cessation

Predictor variables: Participation in cognitive behavioral therapy, Beck Anxiety Inventory Score, and years of education
  Type and level of measurement:
    Smoking cessation success: Categorical (binary)
    Participation in cognitive behavioral therapy: Categorical
    Beck Anxiety Inventory Score: Ratio
    Years of education: Ratio

Sample: 150 participants are recruited from a primary care clinic; each participant selected smokes at least ½ pack of cigarettes per day, and their physician has recommended smoking cessation. All 150 participants have indicated a desire to quit smoking. The 150 participants are selected consecutively from a random sample of charts until the 150 openings are filled. Cognitive behavioral therapy is recommended to all participants by their physician as per the standard of care for that clinic. Six months after each participant is advised to quit smoking and offered cognitive behavioral therapy, they are seen by a research assistant during a regularly scheduled visit with their doctor. The research assistant records their smoking status – smoking cessation success is defined as having no cigarettes for the past one month. The research assistant also administers the Beck Anxiety Inventory, and determines whether the participant participated in cognitive behavioral therapy.

A Logistic Regression Analysis is conducted using the obtained data to see if the data support the idea that smoking cessation success or failure can be predicted by some combination of education, self-reported anxiety level (i.e. as measured by the Beck Anxiety Inventory), and involvement in cognitive behavioral therapy. Fortunately, our data fit all of the assumptions for Logistic Regression Analysis, including that the relationship between success and each Beck Anxiety Inventory Score and Education is linear. Also, the correlations between the three predictor variables are fairly small, meaning there is no multicollinearity. In addition to determining whether the three predictor variables predict success as a whole (the model), the analysis allows us to see which of the predictor variables is significant in adding to the prediction, and in what direction (e.g. participating in therapy is associated with success and not failure, more education is associated with more success, and less anxiety is associated with success?).

5.5.1 Statistical Software Packages for Conducting Logistic Regression

I won't even joke about conducting a Logistic Regression Analysis by hand. Do not pass Go, and skip straight to the statistical analysis package of your choice and be done with it. If you are a member of the SAS team, you can use Proc Logistic. With SPSS, you can select the handy old Analyze tab, then choose "Regression." From the next drop-down menu, you can select "Binary Logistic" (which would be appropriate for the study in Table 5.17) or "Multinomial Logistic." In R, you can use the glm() function. In Stata using the GUI, you will go to the Statistics tab again, select "Binary outcomes" ➔ "Logistic regression, reporting odds ratios" (or "reporting coefficients").

5.6 Discriminant Function Analysis

Discriminant Function Analysis is similar in many ways to logistic regression in that the goal is to predict membership in one level or another of a categorical variable based on a set of predictor variables. Whereas the predictor variables in logistic regression can be continuous or categorical themselves, all the predictor variables (also called attributes) in Discriminant Function Analysis must be measured on an interval or ratio scale. Discriminant Function Analysis also sets a higher bar in terms of assumptions, meaning the assumptions are more difficult to meet, making logistic regression a better choice in terms of ease. Inclusive, Discriminant Function Analysis requires multivariate normality. On the other hand, Discriminant Function Analysis does not require the large sample size required with logistic regression. Finally, Discriminant Function and Logistic Regression Analysis answer slightly different questions. While logistic regression answers the question of likelihood of membership in a particular group, Discriminant Function Analysis tells you which group the individual is more likely to belong to. See Peter Steyn's website for a more thorough discussion of the difference between Discriminant Function and Logistic Regression analyses (Steyn, 2015).

In Table 5.18, I present an example of a study in which Discriminant Function Analysis is employed. This study is conducted to see how accurately we can classify patients as achieving successful smoking cessation versus being unsuccessful in achieving cessation based on a set of variables or characteristics.

5.6.1 Statistical Software Packages for Conducting Discriminant Function Analysis

Again, do not pass Go and skip straight to the old statistical software package of your choice. If your choice is SAS, you will use Proc Discrim. In SPSS you

Table 5.18 An example of a study using Discriminant Function Analysis.

Criterion variable: Smoking cessation
Levels:
  1: Successful cessation
  2: Unsuccessful cessation

Predictor variables: Weeks of cognitive behavioral therapy, Beck Anxiety Inventory score, and years of education
  Type and level of measurement:
    Smoking success: Categorical (binary)
    Weeks of cognitive behavioral therapy: Ratio
    Beck Anxiety Inventory score: Ratio
    Years of education: Ratio

Sample: 50 participants are recruited from a primary care clinic; each participant selected smokes at least ½ pack of cigarettes per day, and their physician has recommended smoking cessation. All 50 participants have indicated a desire to quit smoking. The 50 participants are selected consecutively from a random sample of charts until the 50 openings are filled. Cognitive behavioral therapy is recommended to all participants by their physician as per the standard of care for that clinic. Six months after each participant is advised to quit smoking and offered cognitive behavioral therapy, they are seen by a research assistant during a regularly scheduled visit with their doctor. The research assistant records their smoking status – smoking cessation success is defined as having no cigarettes for the past one month. The research assistant also administers the Beck Anxiety Inventory, and determines how many weeks each participant attended cognitive behavioral therapy.

A Discriminant Function Analysis is conducted using the obtained data to see how accurately we can classify patients as a successful quitter or a nonquitter based on their education, number of sessions of cognitive behavioral therapy, and their anxiety level. Thankfully, our data meet all the required assumptions for Discriminant Function Analysis, including the absence of outliers and multicollinearity, and the variance/covariance matrices are homogenous. Thus, we can interpret our findings with confidence.

---

will select the Analyze tab yet again, and choose Classify ➔ Discriminant. In R, you will want to use the lda() function for a linear discriminant analysis. Finally, in Stata using the GUI, you will again select your Statistic tab, choose "Multivariate analysis" ➔ "Discriminant analysis" ➔ "Linear (LDA)."

5.7 Multiple Comparisons

Before I move on to the topic of nonparametric statistics, I want to reiterate the concern about multiple comparisons. Ironically, I have made this point "multiple" times in this primer, but it cannot be overstated – the more comparisons we make using the same set of data, the more we risk making a Type I error. Sometimes the increased risk of Type I errors is referred to as the family-wise error rate.

The risk of making a Type I error increases because there is a risk of making a Type I error every time we compare a statistic to a critical value associated with a particular α, rejecting the null hypothesis when our statistic exceeds that critical value. If our p value, α level, is set at 0.05, we have a 5% chance of making a Type I error; that is, rejecting a null hypothesis when the null hypothesis is true. If we run two tests, or comparisons in the case of a t or F test, we have a 10% chance of making at least one Type I error ($0.05 + 0.05 = 0.10$). If we make three comparisons, our risk is 15%, and so on.

As noted previously in this chapter, we can protect against the increased Type I error rate by adjusting our α, or by making adjustments to the way the comparisons are conducted. Designated post hoc and planned comparisons reduce the increased risk of making Type I errors in this way. There are various options for running post hoc and planned comparisons, and each has advantages and disadvantages.

Generally, planned or post hoc comparisons are made after we have conducted one of the NOVAs and we obtain a significant F. The F test can be thought of as an omnibus test – it tells us that there are differences between group means somewhere in our data, but when there are more than two means being compared with the F test (such as with an independent variable with three or more levels), we don't really know which group(s) differed from the others. Therefore, we may want to analyze our data further, either with pre-determined contrasts or by exploring our data and comparing all possible groupings.

When running two or three planned comparisons, or contrasts, it may be feasible to use a Bonferroni correction, which is simple to calculate and easy to interpret and explain. The Bonferroni correction adjusts the α level that we use to reject the null hypothesis of no difference. The correction is calculated by simply dividing the set α level by the number of comparisons we are making. We can also compare means of combined groups to single or combined group means; the latter are called complex comparisons. The drawback of the Bonferroni correction method is that it is very conservative, or strict; if we have more than a few comparisons, we will set α exceedingly low, thus decreasing our power substantially.

If we are only interested in comparing our control group to each of our experimental groups, and we do not need to compare each experimental group to one another, we can use the Dunnett test. This option also reduces the number of comparisons based on pre-existing ideas.

When we want to go ahead and explore our data, analyzing comparisons between all possible group pairings, we can use the Least Significant Difference (LSD) test, which is based on comparing all of the differences between groups to the smallest difference between means that is significant when conducting a t-test. The Type I error rate tends to be elevated with the LSD. Another option is the Tukey Honestly Significant Difference (Tukey HSD) test, which adjusts the alpha level based on the number of comparisons to maintain a level Type I error rate. The Newman–Keuls test similarly allows us to compare pairs of means, but this is done sequentially starting with the two groups with the largest difference between their means and continuing to the next largest mean difference, and so on until the difference between a pair of means is no longer significant. This has the effect of reducing the number of comparisons. Hervé Abdi and Lynne Williams (2010) provide a good discussion of the difference between the Tukey and the Newman–Keuls, both of which are based on the Studentized range (i.e. the range of means divided by the estimated standard error of the mean; see Stats Direct (2015–2018b). Finally, when we want to make multiple comparisons of pairs of groups as well as comparisons between the means of combined groups, we can use Scheffe's test. Penn State's Department of Statistics provides an exceptional description and comparison of all of the major planned and post hoc comparisons (Penn State University, 2018).

It should be noted that some researchers skip the ANOVA altogether and go straight to t-tests. If they do this, and conduct multiple t tests, they are elevating their risk of making a Type I error. They should either adjust their α level accordingly to protect against the elevated risk, or they should start with an ANOVA and, only if significant, follow-up with the appropriate planned or post hoc comparisons.

5.8 Summary

In this chapter, I have reviewed a set of statistical analyses that are used fairly commonly in clinical and educational research. There are many more statistical analyses than what I have laid out above, including that I have not discussed some of the more complex (to me) time series techniques used for forecasting, path analyses for testing hypothesized causality between variables, or structural equation modeling, another tool for evaluating causality between variables. These topics are reviewed in more advanced textbooks, and there are some good online resources as well. If you are inclined to use any of these techniques, I would recommend consulting with an expert, possibly completing a course (with some that are available online), and/or obtaining and reading a good tome.

I have also not reviewed factor analytic techniques or cluster analysis, great tools for reducing data by combining closely related variables into units, such as factors or clusters. These new variables (factors, clusters) can be analyzed using the appropriate statistical technique, but now with fewer variables. There are a lot of good resources for learning more about factor and cluster analysis, including the online manuals for statistical software packages.

I have attempted to cover some of the basic parametric statistical tests that you are likely to come across in reading others' clinical and educational research. Ideally, you will now be able to read research papers using the statistical tests reviewed in this chapter and have at least a basic understanding of the meaning and purpose of the tests. Hopefully, these will also be helpful in designing your own studies if you do your own research as well.

In the next chapter, I review the nonparametric relatives of many of the tests that I have included in this chapter. Nonparametric statistical tests are used when our data do not meet the assumptions required for parametric analyses.

6

Nonparametric Statistical Techniques

So far, we have talked about some of the more commonly used parametric statistical techniques for analyzing research data. As you recall, all of these statistics have certain requirements or assumptions. There will be times, though, when we cannot meet the assumptions. At these times, we can analyze our data using nonparametric statistical tests.

Nonparametric statistical tests can be used when our data are not normally distributed, when we have outliers, when we have major differences in group variances, and/or when we have ordinal or frequency data. Frequency data are represented by counts of members of a category – e.g. the number of females and the number of males, the number of individuals with each blood type, or the number of individuals who achieve remission from a disease. Nonparametric tests are also useful when we have values that cannot be fully measured to the end points – such as survival duration when some individuals have ongoing survival, or HIV viral load that, at one end of the distribution, can only be quantified as "not detected." In those instances, nonparametric tests that are reliant on medians, ranks, or frequencies can be utilized. A median, as discussed in Chapter 5, is the middle score, or the midpoint in a range of scores among a group of observations or individuals.

The advantages of nonparametric tests include their freedom from most of the assumptions of parametric tests as well as the relative ease with which they can be conducted. The drawback is that they tend to have less power than parametric tests. Thus, you may be less likely to reject a null hypothesis when, in fact, the null hypothesis is not true (Type II error). In other words, you may miss out on being able to demonstrate significant findings consistent with your prediction. This would not be due to your prediction or hypothesis being incorrect, but simply due to not having a statistical test sensitive enough to "prove" your hypothesis.

Research Design and Analysis: A Primer for the Non-Statistician, First Edition. Leslie D. Rosenstein.
© 2019 John Wiley & Sons, Inc. Published 2019 by John Wiley & Sons, Inc.

6.1 Chi-Square

Chi-square (χ^2) is a statistical test we can use when we have frequency data. It works by comparing observed to expected frequencies. It can be used to compare the relative frequencies of one outcome versus other outcomes across groups. It can also be used to test for an association between one grouping and another (Chi-square test of independence). Finally, Chi-square can be used to test whether the observed frequencies are consistent with expectation based on theory or known population frequencies (goodness of fit).

A statistically significant Chi-square test indicates that the observed frequencies are significantly different than the expected frequencies. In the case of a goodness of fit test, for example, this would indicate that our data do not fit with expectation based on a known or comparison population and, thus, our sample apparently does not come from the comparison population.

Happily, a Chi-square test can be calculated by hand fairly easily, with a statistical software package, or with one of several handy and simple-to-use online calculators that can be found with a quick online search (to be reviewed further in Section 6.1.1). There are two major criteria for being able to conduct a Chi-square test. As noted, the data must be frequency in nature. The second criterion is that the *expected* value of any particular subgroup generally cannot be fewer than five observations. The concept of expected values will become clearer (hopefully) as I describe the calculations for Chi-square.

To make this discussion less abstract, let's use an example of a 2×2 contingency table with hand preference and gender as our two dichotomous variables (see Table 6.1). We can use the data in our table to calculate whether there is an association between gender and hand preference by comparing the observed frequencies to the frequencies that would be *expected* if there was no difference in hand preference across the genders. In this case, the null hypothesis would be that there is no difference in hand preference across the genders; that is, there is no association between hand preference and gender. The alternative hypothesis, based on the known literature on hand preference, would likely be that there are relatively more left-handers among the males than among the females. This would be supported by a significant Chi-square (denoted by the

Table 6.1 Hypothetical 2×2 contingency table for hand preference and gender.

	Right-handed	Left-handed	Total
Male	178	22	200
Female	191	9	200
Total	369	31	400

$\chi^2 = 5.91, p = 0.015.$

Greek symbol, χ^2) with proportionately more left-handers among the males than the females (see Table 6.1).

The Chi-square is calculated by comparing the observed and expected frequencies. We would start by calculating the expected frequencies for each cell by multiplying the corresponding row total by the corresponding column total and dividing that number by the grand total. So, for instance, in the contingency table in Table 6.1, we would multiply the total number of right handers (369) by the total number of males (200), and then divide that number by the grand total of individuals (400). We would calculate the expected frequency for each cell in the same manner.

In the case in which we are simply testing whether our groups are evenly distributed, as shown in Table 6.2, our expected frequencies would simply be equal portions of the grand total of the sample. As can be seen in Table 6.2, the grand total of the three groups is 90. So, in the event that color preference is evenly distributed (i.e. the null hypothesis that there is no color preference in this sample), the expected frequency for each cell would be 30. In this example, the overall sample actually does show a color preference for burnt orange or maroon relative to green. Green was the least likely to be chosen as a favorite color.

Once we have calculated the expected frequencies for a Chi-square, the rest of the computation is simple. Each of the expected values is subtracted from the respective observed value of the respective cell. Those differences are squared and then divided by the expected frequency. Finally, the resulting numbers are summed and, voila, we have our χ^2. This might make more sense if you see it in a formula (see Table 6.3).

While the nonparametric tests have less stringent assumptions than the parametric statistical tests, they do have some assumptions or limitations,

Table 6.2 Sample Chi-square table comparing groups.

Favorite color of club members	Observed frequency	Expected frequency
Burnt orange	38	30
Maroon	37	30
Green	15	30
Total	90	90

$\chi^2 = 11.27$, $p = 0.004$ (therefore, the group shows significant color preferences).

Table 6.3 Formula and sample calculation for χ^2.

$\chi^2 = \Sigma(o-e)^2/e$ From Table 6.2: $\chi^2 = (38-30)^2/30 + (37-30)^2/30 + (15-30)^2/30 = 11.27$

nonetheless. The assumptions for Chi-square as thoroughly outlined by Dr. Mary McHugh (2013) and others are:

1. The data must be frequency data – they must be counts of an event or occurrence.
2. The categories must be mutually exclusive such that an individual may only fit into one cell, that is, one level of each variable (i.e. an individual cannot be both right-handed and left-handed, they must be either one or the other for a Chi-square test).
3. Somewhat related, there are no repeated measures – the same individual does not appear in the table more than once due to repeated assessment.
4. Observations are independent – individuals are unrelated, unpaired.
5. As noted earlier, the expected frequency of any given condition should be at least five, though there may be exceptions to this with larger numbers of cells.
6. And, as noted by Young and Veldman (1981), the sum of observed frequencies in the Chi-square table must equal the sum of expected frequencies.

6.1.1 Statistical Software Packages for Conducting Chi-Square

As mentioned earlier in this chapter, there are several simple-to-use online calculators for computing the Chi-square statistic. Just use your favorite search engine and search for, "Chi-square online calculator" and you will get a number of results returned. These calculators typically have a simple fill-in-the-blank form that lets you fill in cell values (frequencies) and, in some cases, even the variable labels. If you have trust issues, and feel uncertain about the results using an online calculator, it is quite easy to pull up more than one calculator and try each to make sure you obtain the same result.

If you already have a data set with the respective data for which you wish to run a Chi-square, then it is also easy to perform the test using your preferred statistical software package. In SAS®, you would use the Proc Freq procedure (as in frequency) with the Tables statement and the Chisq option. In Appendix B of this primer, I present a sample, fictional study analyzed using SAS (see Table B.6).

If you prefer SPSS®, you would go to the Analyze tab, choose "Descriptive Statistics," and then "Crosstabs." In R, one option would be to use the chisq.test function after you create a table using the "tbl" function. In Stata® you would go back to your Statistics tab, select, "Summaries, tables, and tests" → "Tables" or "frequency tables" → "Two-way tables with measures of association" or "two-way measures of association" → "Pearson's Chi-squared." You can also use the Excel spreadsheet. Under the "Formula tab," select "More Functions" → "Statistical" → "CHISQ.TEST."

6.2 Median Test

Mood's median test is a special case of Chi-square test that analyzes whether the medians of two or more samples are equal. This test is especially handy when we have outliers that skew our means substantially, making t tests and ANOVA inappropriate. In the presence of outliers, the Median test is also better than the Kruskal–Wallis test, which I review further on in this chapter. It is also handy when we have ordinal data. For instance, we can compare different ethnic groups and their placement on a transplant list in order to see if one group is somehow favored over the others (all other things being equal), or vice versa. It is a special case of Chi-square because it compares the frequencies of scores above versus below the grand median across groups. In other words, we compare groups to one another by considering how many scores, or ranks, in each group are above or below the overall median score when all the groups are combined. In Table 6.4, you can see an example of a study in which the Median test can be applied. Notice that this study is a variation of the Analysis of Variance example from Chapter 5.

6.2.1 Statistical Software Packages for Conducting Median Tests

If you wish to use a statistical software package to conduct a Median test, you have multiple options. With SAS, you can use the Proc NPAR1WAY statement with the option of Median. In SPSS, you will go to your Analyze tab and select, "Nonparametric Tests" → "Legacy Dialogs" → "K̲ Independent Samples" → "Median." In R, you can use the Mood.medtest function followed by the pairwiseMedianTest function (and, yes, that is the correct spelling and punctuation). In Stata, using the graphical user interface (GUI), you can go back to your Statistics tab, choose "Nonparametric analysis" → "Tests of hypotheses" → "K-sample equality-of-medians test."

6.3 Phi Coefficient

The Phi coefficient is a measure of the relationship between two dichotomous variables. A dichotomous variable is a nominal or categorical variable with two levels (such as male and female, left- and right-handed). Since we use the frequency of occurrence in each level of the dichotomous variables, we rely on the Chi-square test to calculate the Phi coefficient. However, we take the Chi-square test of association between our two dichotomous variables one step further by dividing the Chi-square value by the size of the sample and then calculating the square root of that quotient ($\phi = \sqrt{\chi 2 / N}$). This gives us a value between −1 and +1, indicating the strength of the association between the two variables, much like r in the correlation between two continuous variables.

Table 6.4 Example of a study using the median test.

Independent variable: Treatment

Levels:

    1: New adjunct medication

    2: Cognitive behavioral therapy (CBT)

    3: New adjunct medication + CBT

    4: Care as usual

Dependent variable: Time to recurrence

Sample: All participants are patients in an outpatient psychiatry clinic diagnosed with major depression:

50 are treated with the addition of a new medication

50 are treated with CBT

50 are treated with the addition of a new medication plus CBT

50 are treated with their care as usual

Each patient is randomly assigned to either the control (care as usual) group, the new adjunct medication group, the CBT group, or the combined medication plus CBT group after consenting to be in the study. Their records are reviewed three years later by a research assistant who records if the patient ever achieved remission, and when they are reported to have experienced a recurrence of their depression. If the patient never achieves remission, they are considered to have a recurrence at 0 weeks or less. If the patient remains in continued remission at the time of the three-year follow-up, they are considered to have a recurrence at >156 weeks.

A Median test is performed to determine whether the median time to recurrence differs across the four groups. The prediction is that the result will be significant, with the care as usual group having the shortest remission as demonstrated by a lower median time to recurrence. Pairwise Median tests can also be calculated to compare one group to another, such as to see if patients receiving CBT fared better or worse than patients receiving the medication without CBT. When multiple comparisons are made in this manner, the *p* value will have to be adjusted to account for the increased risk of a Type I error as discussed in the section toward the end of Chapter 5 regarding multiple comparisons.

When we have more than two levels of each categorical variable, we can use Cramer's V in place of the Phi coefficient.

We could use the same data and setup as in Table 6.1 (the gender and hand preference example) to calculate Phi – whereas the Chi-square test showed us that there is a significant association between gender and hand preference, the Phi coefficient will give us an idea of the *degree* or *strength* of the relationship between the two dichotomous variables. In that regard, Phi can be thought of as a measure of effect size.

In Table 6.5, I provide a different example of a study in which the Phi coefficient could be applied. Here, we are considering the strength of the relationship between participation in cognitive behavioral therapy and the risk of suicide attempt.

Table 6.5 Example of a study using Phi.

Variables: Participation in cognitive behavioral therapy (CBT) and suicide attempt
    Type and level of measurement:
        Participation in CBT, Yes or No: Categorical, Frequency
        Suicide attempt, Yes or No: Categorical, Frequency
Sample: Records of patients diagnosed with major depression in an outpatient psychiatry clinic are reviewed. In total, there are 200 records. A research assistant records whether a patient participated in CBT (yes or no) in the past two years, and whether they had a suicide attempt during that time (yes or no).
A Phi coefficient is calculated between the CBT status and suicide attempt status. A significant, negative relationship would tell us that participation in CBT is associated with less likelihood of a suicide attempt (or vice versa). As another reminder, a significant relationship does not prove causality, only an association. There may be a third, unrelated variable contributing to the relationship. Even if there is a causal relationship, this study design would not allow us to determine the direction of causality. For instance, having a suicide attempt may have decreased the opportunity to participate in therapy, *or* being in therapy may have helped prevent a suicide attempt; either could be argued based on this study's data.

6.3.1 Statistical Software Packages for Calculating the Phi Coefficient

As with the Chi-square test, you can calculate the Phi coefficient by hand. Or, you can easily find simple Phi coefficient calculators online with a quick search using your preferred search tool. With SAS, you again turn to Proc Freq and enter the Chisq option. When you use SAS in this manner to calculate Chi-square, the Phi coefficient is also returned in the Results. In R, you can use the phi function. With Stata, you can go back to the Statistics tab, select, "Summaries, tables, and tests" → "Tables" or "frequency tables" → "Two-way tables with measures of association" or "two-way measures of association" → "Cramer's V." In case you are confused by the choice of Cramer's V, you should know that with a 2×2 contingency table (i.e. two levels of the two variables), Cramer's V = Phi.

6.4 Mann–Whitney *U* Test (Wilcoxon Rank Sum Test)

The Mann–Whitney *U* Test is most similar to a *t* test for independent samples, and it is used when the assumptions of the *t* test are not met and the samples are small. The test is based on rankings, so it can also be used with ordinal data. Here, instead of comparing the average raw score between two groups, the scores are converted to ranks. The ranks are based on the total sample. The two groups, then, are compared to determine if one group has higher overall rankings than the other. The Mann–Whitney *U* test can be calculated by hand, with an online calculator, or with a statistical software package. It is also sometimes referred to as the Wilcoxon rank sum test.

Unlike most other statistics, in the Mann–Whitney U test, a smaller value of U is associated with rejecting the null hypothesis. That is because U tells you how much overlap there is between the two groups in their rankings. If there is no overlap at all, then $U = 0$. That means that all observations in one sample are ranked below every one of the observations in the second sample. The more overlap between the two groups' ranks, the larger the size of U. So, with the Mann–Whitney U, we are looking to come in under the critical value in order to reject the null hypothesis that there is no significant difference in the rankings of the two groups.

In Table 6.6, I offer an example of a study in which the Mann–Whitney U test is used. You will probably notice an astonishing similarity to the study in Table 5.6 when I was reviewing t tests.

As mentioned in Table 6.6, the data in our depression remission study are likely to be positively skewed. This means that, instead of the data having the appearance of a bell-shaped curve when plotted on a frequency polygon, the data will have an asymmetric distribution with the long "tail" to the right. That is because remission can only be so short (e.g. a couple of days in a worst-case scenario), but it can extend to infinity. So, if most people (the mode or the top

Table 6.6 Example of a study using a Mann–Whitney U test.

Independent variable: Treatment
Levels:
1: New adjunct medication
2: Care as usual
Dependent variable: Number of weeks in remission from depression
Sample: All participants are recruited from an outpatient psychiatry clinic where they have previously been diagnosed with recurrent major depression, but have had resolution of their symptoms (remission) within the past month:
50 are treated with their care as usual
50 are treated with a new medication developed to prevent relapse of depression
Each patient is randomly assigned to either the control (care as usual) group or the new medication group after consenting to be in the study. After two years, a research assistant reviews the records to determine when patients had a recurrence of depression, if at all yet – this is recorded as weeks in remission.
Because our data will likely be positively skewed (see Figure 6.1), the data do not meet the assumption of normality required for a t test. Also, some patients can only be characterized as indeterminate at the positive end of the distribution (because they are still in remission at the study end date). For these reasons, we really cannot do a parametric t test. Instead, a Mann–Whitney U test is calculated in order to determine whether the two groups differ in the duration of remission – our hypothesis is that those taking the new medication will maintain remission longer than the care as usual group.

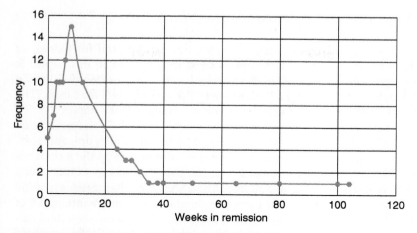

Figure 6.1 Illustration of a positively skewed distribution.

of the distribution) remain in remission about one year, the distribution will be asymmetric because it can only go so far to the left (from one year down to zero weeks), but it can extend up, to the right of the mode, from one year to infinity. Thus, it can have the appearance of the distribution illustrated in Figure 6.1. In Figure 6.1, the modal duration of remission is shown as about 10 weeks with one person in remission at 2 years.

6.4.1 Statistical Software Packages for Conducting a Mann–Whitney U Test

Believe it or not, there are even online calculators for computing a Mann–Whitney U test. Again, a quick search using your favorite search engine will pull up several options, and they are quite simple to use. If you already have your data included as part of a larger data set, though, you can also use your preferred statistical software package. As with the Median test, when using SAS you will use the Proc NPAR1WAY statement with the option of "Wilcoxon rank-sum" (which is another name for Mann–Whitney U). In SPSS, you again go to your Analyze tab, select "Nonparametric Tests" → "Legacy Dialogs" → "2 Independent Samples," and then in the dialog box check the box by "Mann–Whitney U" under "Test Types." With R, you will use the wilcox.test() function. In Stata, using the GUI, you can go back to your Statistics tab, again choose "Nonparametric analysis" → "Tests of hypotheses" and this time select, "Mann–Whitney two-sample ranksum test." One note, these commands will vary slightly depending on the edition of Stata that you are using, but the changes from edition to edition seem to be minor.

6.5   Sign Test and Wilcoxon Signed-rank Test

The sign tests are also analogous to *t* tests, but can be used for paired, or matched, groups, like the matched pairs *t* test that I reviewed in Chapter 5. These nonparametric versions of the matched pairs *t* test are beneficial when we have outliers or other major violations of the assumptions for *t* tests. The sign test can also be used for single samples and independent samples.

The calculation of the sign test is pretty straight forward if you have a good grasp of the binomial distribution. For example, when analyzing pre- and post-test scores using the matched-groups format, we simply add up the number of positive change scores (i.e. successes) and compare that to expectation using the binomial distribution. The null hypothesis would be that there is a 50/50 chance of success. In the case of a study evaluating a new intervention, if 9 of 10 people have higher post-intervention scores (i.e. successes), we would calculate the probability of that happening by chance alone. Specifically, we would calculate the probability of there being either 9 *or* 10 positive change scores given a probability of positive change of 0.5 (the null hypothesis). We would add the probability of each (9 and 10 positive change scores). In this example, the probability of obtaining 9 or more positive change scores in the face of chance is 0.01. In this case, we would reject the null hypothesis that there were not more positive changes, or successes, than chance.

Note that there are plenty of binomial calculators, as well as binomial tables, available online. The sign test can also be conducted using a statistical software package (to be discussed in Section 6.5.1).

Whereas the sign test only analyzes whether the signs (plus or minus) between groups are different, the Wilcoxon signed-rank test additionally analyzes the magnitude of the difference by considering ranks as well as signs. To be able to run a Wilcoxon signed-rank test, though, the distribution of the differences between our scores or ranks must be symmetrical (though you may be able to transform your difference scores using a mathematical technique in order to get a symmetrical distribution). The Wilcoxon signed-rank test can be calculated by hand or with a statistical software package.

In Table 6.7, I have an example of a study that can be analyzed with a Wilcoxon signed-rank test. In this example, I have a repeated measures design, much like that of a matched pairs *t* test. My point of interest is whether patients experience a decrease in suicidal ideation with a new medication as compared to a placebo. I am using the nonparametric test instead of the *t* test because the data are not normally distributed. This is in part due to some restriction of range, but also due to some outliers with frequent suicidal ideation that cause the distribution of "scores" to be positively skewed. Fortunately, in our study, the distribution of the difference scores is symmetric, and not skewed.

Table 6.7 Example of a Wilcoxon signed-rank test.

Independent variable: Treatment
  Levels:
    1: New medication
    2: Placebo
Dependent variable: Number of days with suicidal ideation

Sample: 50 participants are recruited from an outpatient psychiatry clinic where they have been diagnosed with chronic, severe major depression with suicidal ideation. Half the patients receive the new medication first for four weeks, followed by the placebo for four weeks, and half receive the placebo treatment first followed by the medication.

Patients and their providers are blind to whether the patient is taking the placebo or the new medication throughout the study period. The participants keep a daily diary recording whether they experienced suicidal ideation during each day. A research assistant reviews the diaries after each four-week period and records the number of days with suicidal ideation into the data set along with each respective participant's status at the time as being in the placebo phase of treatment or the medication phase. Our hypothesis is that there is a difference between the two conditions, and that the new medication is associated with a decrease in days with suicidal ideation. A Wilcoxon signed-rank test is performed, with a significant result indicating that there is a significant difference in the number of days with suicidal ideation with the new medication versus the placebo. Inspection of the median days with suicidal ideation in the two conditions will indicate whether there are fewer or more days of suicidal ideation with the new treatment versus the control.

6.5.1 Statistical Software Packages for Conducting Sign Tests

Yep! Once again, there are easy-to-locate online calculators for computing a sign test and the Wilcoxon signed-rank test. If your data are large or part of a broader data set, you can also use statistical software packages. With SAS, you will go back to Proc Univariate, which returns the Student's t (which can be ignored in this instance) as well as the Sign test and the "Signed Rank." When you have a repeated measure like the example shown in Table 6.7, you will first create a new variable in the data step. This new variable will be the difference between the two scores for each participant/individual. In SPSS, you will go to your Analyze tab and again select, "Nonparametric Tests" → "Legacy Dialogs" → "2 Related Samples." In the dialog box that opens next, you will check "Wilcoxon" or "Sign" under "Test Type." In R, you will use the wilcox.test or the SIGN.test function. In the cased of paired data, as in the example shown in Table 6.7, you will specify that the values are paired (TRUE). In Stata, you visit the Statistics tab again, choose "Nonparametric Analysis" → "Tests of hypotheses" → "Wilcoxon matched-pairs signed-rank test" or "Tests of equality of matched pairs."

6.6   Kruskal–Wallis Test

The Kruskal–Wallis test is sometimes referred to as a one-way ANOVA for ranks. It is used to compare two or more independent samples when our data violate the assumption of normality and our sample size is not large, or when we have ordinal data. If our data can be transformed to a normal distribution and we have interval data, we should stick with the parametric ANOVA given its better power.

While ANOVA tests for differences in raw scores across groups, the Kruskal–Wallis analysis tests for differences in grand ranking across groups. That is, each raw score is converted to a rank based on the combined groups, and then the rankings are compared between the groups. The resulting statistic in the Kruskal–Wallis test is *H*. As with the other nonparametric tests reviewed earlier in this chapter, the Kruskal–Wallis test can be calculated by hand, with an online calculator, or with a statistical software package.

As noted above, with Kruskal–Wallis, we do not have to assume our data are normally distributed. However, the distributions of the scores in each group must be the same if we wish to compare medians, and the variances of each group should be approximately equal. If the shapes of our distributions are different, or we have extreme outliers, we can revert to using the Median test (discussed in Section 6.2). The Median test is more robust to outliers than the Kruskal–Wallis test. Unfortunately, the Median test is not as powerful as the Kruskal–Wallis test. For a good discussion of when to use the Kruskal–Wallis test, I would recommend a quick reading of *Kruskal–Wallis ANOVA: Use & Misuse* (Dansfield & Brightwell, 2013) and *Data Considerations for Kruskal–Wallis Test* (Minitab Express Support, 2017).

If you noticed that the Kruskal–Wallis test and the Median test seem similar, you are right. There are certain advantages of each, however. The Median test has fewer restrictions in terms of assumptions and will be better when data are particularly skewed. On the other hand, it is less precise, and thus less powerful, than the Kruskal–Wallis test in that it only considers the number of observations above or below the grand median across groups.

In Table 6.8, I present an example of a study that can be analyzed with a Kruskal–Wallis test. Here, I am mixing it up a little with a scenario conducive to using ordinal data, for which the Kruskal–Wallis test is well suited. This will also be a refreshing change for readers more interested in educational topics.

6.6.1   Statistical Software Packages for Conducting a Kruskal–Wallis Test

And, another shout out for the World Wide Web: once again you can find handy, easy-to-use online calculators to perform a Kruskal–Wallis test. There are also options for using statistical software packages. With SAS, we go back

Table 6.8 Example of a Kruskal–Wallis test.

Independent variable: Study skills training
    Levels:
        1: No training
        2: Study skills handouts
        3: Study skills class

Dependent variable: Relative class rank at graduation

Sample: 45 participants are new freshman at a midsized high school. Exclusionary criteria are a diagnosis of a learning disorder, ADHD, or intellectual disability:
    15 matriculate in the traditional curriculum
    15 are additionally provided with handouts about study skills during orientation
    15 are enrolled in a first-semester class devoted to teaching study skills using a structured curriculum

Each student is randomly selected from the freshman class and randomly assigned to either the control (traditional curriculum) group, the new study skills handouts group, or the study skills class group. At the end of four school years when the students graduate, the relative class ranks among the 45 are recorded into the data set. A Kruskal–Wallis test is computed to determine if the relative class ranks differ between the three groups. The hypothesis is that there will be differences between the groups, with those completing the study skills class achieving the highest relative class ranks. We also obtain Dunn's post hoc test to make pairwise comparisons between the three groups. Our hypothesis is that all three groups differ from one another, with the study skills class group outperforming both of the other two groups, and the handouts group outperforming the control group.

to Proc NPAR1WAY and use the Wilcoxon option again. When performing a Wilcoxon rank sum test (aka, a Mann–Whitney U test), the Kruskal–Wallis test is automatically performed when there is more than one level of the independent variable. In SPSS, you will go back to the Analyze tab, and then you can choose "Nonparametric Tests" → "Legacy Dialogs" → "K̲ Independent Samples" and then, in the dialog box that opens, check "Kruskal–Wallis H" under "Test Type." In R, you can use the Kruskal.test function. In Stata, you will go to the Statistics tab, choose "Nonparametric Analysis" → "Tests of hypotheses" → "Kruskal–Wallis rank test."

6.7 Rank-Order Correlation

In Chapter 5, we talked about correlation, referring to the Pearson correlation between two continuous variables, and I also mentioned the point–biserial correlation between one continuous and one dichotomous variable. Those correlational analyses, like the other parametric statistics, require certain assumptions to be met. When those assumptions are not met, the alternatives are Spearman's rank-order correlation, Kendall's tau, and gamma.

Spearman (ρ, pronounced, "rho") is similar to the Pearson r, but it is based on the relationship between the rankings of two variables. It is useful for measuring the strength and direction of the relationship between two variables when at least one of our variables is measured as ordinal data (e.g. placement on the transplant list or severity of illness as measured on a Likert scale), when our data violate the assumptions of normality, or when we have a curvilinear relationship between our variables. While the relationship between the two variables in Spearman's rank-order correlation does not have to be linear, we still expect at least a monotonic relationship; that is, as the value of one variable increases, the value of the second variable increases or decreases, just not necessarily at a uniform rate. In other words, Spearman's rank-order correlation would not be suited to pairs of variables with a u-shaped or inverted u-shaped relationship (such as when high and low values of one variable are associated with high values of the other variable, but midranged values of the first variable are associated with low values of that second variable).

Kendall's tau also measures the association between two sets of rankings, and it has the same assumptions as Spearman's rank-order correlation. Tau is a better option than Spearman's rho when we have smaller samples, when we have many ties in our scores/ranks, or when we are concerned about errors in the data coding. Tau is based on concordance and discordance between pairs of scores, so it really tells us more about the agreement in ranks. There are actually three versions of tau (a, b, and c) depending on certain technical characteristics of the data, though a discussion of those variations is beyond the scope of this primer (and my brain's central processing unit, CPU).

A third, less well-known measure of association for ranked data is Goodman and Kruskal's gamma. The gamma coefficient is useful when you have ordinal variables and a lot of ties in your data; that is, when many observations have the same score and, thus, the same rank on a variable.

In Table 6.9, I offer an example of a study in which a Spearman rank-order correlation would be helpful to calculate. Here, I go back to the educational scenario, but now we are interested in the relationship between actual time spent studying (and doing homework) on a weekly basis and relative class rank at graduation. I am particularly interested in the strength of the relationship along with the direction (e.g. I hypothesize that more time spent studying and doing homework is associated with a better class rank, and significantly so).

6.7.1 Statistical Software Packages for Conducting Rank-order Correlations

I am starting to feel like a broken record here, but, yes, there are online calculators available to calculate a Spearman rank-order correlation coefficient. In SAS, you will use Proc Corr again, but this time add the option, "SPEARMAN" or even KENDALL if you need to compute a Kendall's tau (in this case the "b"

Table 6.9 Example of a study with Spearman's rank-order correlation.

Variables: Time spent studying and/or doing homework and relative class rank at graduation

Type and level of measurement:

Time spent studying and doing homework: Ratio

Relative class rank: Ordinal

Sample: 75 students at a mid- to large-sized school are selected at random at the beginning of their freshman year and recruited to participate in the study. Exclusionary criteria are previous diagnosis with a learning disorder, ADHD, or intellectual disability.

Students keep a daily diary documenting the time they spend studying or doing homework. A researcher tallies the total time spent studying and/or doing homework for each participant across the four years of high school. The weary research assistant also records each student's relative class rank among the group of 75 at graduation. A Spearman rank-order correlation is calculated between hours spent studying (including doing homework) and relative class rank at graduation. The result will tell us if there is a significant relationship between the amount of time a student spends studying and their relative class rank. A significant relationship would tell us that time spent studying is *associated with* class rank; as a reminder, a correlation does not prove causality, only an association, or relationship. There may be a third variable contributing to the relationship. For instance, smarter students may be more achievement oriented and driven to study more, and, in turn, smarter students may have higher ranks because of their higher aptitude to begin with.

version of Kendall's tau is returned in the results). In SPSS, you will go back to the Analyze tab, choose "Correlate" → "Bivariate..." and then in the ensuing Bivariate Correlations dialog box that opens, check "Spearman" or "Kendall's tau-b" under "Correlation Coefficients". In R, you will use the "cor.test" function, and specify "method = 'spearman.'" In Stata, you will go to the Statistics tab, choose "Nonparametric Analysis" → "Tests of hypotheses" → "Spearman's rank correlation."

6.8 Summary

Hopefully, I have now given you a decent overview of nonparametric options for statistical analyses. Sometimes, I think we neglect nonparametric statistics a little, or at least I myself have in the past. These tests, while not as powerful as parametric statistical tests, can be quite useful in the right situations. As discussed earlier, they are often necessary because the real world does not always cooperate with us in meeting the assumptions of parametric statistics. Sometimes, we are more interested in things like rank and frequencies, and then the nonparametric tests are especially important. In the next chapter, I review meta-analytic studies. These are technical analyses of cumulative research in the literature, and that research may have been completed with either parametric or nonparametric analyses.

7

Meta-Analytic Studies

The purpose of this chapter is to introduce you to the concept of meta-analytic studies so that you can read published meta-analyses and evaluate them with a critical, though appreciative, eye. While meta-analytic studies can be flawed just as any study can be, these studies have the advantage of including multiple researchers' data and findings across multiple sites. They also require an incredible amount of patience and due diligence on the part of the authors. If you ever consider conducting a meta-analysis yourself, I strongly recommend that you obtain some good reviews of the procedures involved. A couple good ones I have found are: *Meta-analysis in Medical Research* by Haidich (2010) and *Meta-analysis of Clinical Trials* by Sun, Freeman, and Natanson (2018). It would also be extremely helpful to consult with someone who has conducted meta-analytic studies or a statistical consultant with a background in meta-analysis.

A meta-analytic study is typically conducted in order to compile results across studies asking a specific question. Ultimately, the researcher is hoping to obtain a summary effect size that represents the true effect in the population being investigated. Sometimes, there may be a single population to which we want to generalize, while other times there may be more than one population (e.g. patients of different age groups, with different comorbidities, etc.). When that occurs, but the researcher is still interested in the effect of, say, a drug, across those studies, a slight variation of the analysis will be performed, using a random effects model instead of a fixed effects model. Whether we have a fixed effects model (with a uniform population across studies) or a random effects model (with differences across studies in the populations) will impact how each study's effect size will be weighted in calculating the summary effect size (Borenstein, Hedges, & Rothstein, 2007; Hoffman, 2015).

Once the researcher has defined the question, they will select studies and determine how the effect size will be measured. A researcher may choose to limit their studies to true experiments with control groups and random assignment. They may also limit their study selection to studies that include specific data or scores in their papers.

Research Design and Analysis: A Primer for the Non-Statistician, First Edition. Leslie D. Rosenstein.
© 2019 John Wiley & Sons, Inc. Published 2019 by John Wiley & Sons, Inc.

Regardless of their selection criteria, the researcher(s) will then need to do their best to ensure they perform a thorough review of studies conducted that meet those criteria and address the research question. The process for culling the literature should be predetermined and as broad as possible in order to avoid bias. The researcher(s) may use specific and multiple search engines and databases, followed by a review of references within any papers found through the initial search.

Once the studies are selected, the effect size of each study is recorded or calculated, if needed, based on the data reported in the papers. In case you have already blocked them from your mind, I reviewed effect sizes in Chapter 5. Which effect size a researcher chooses to use for their meta-analysis will depend on several factors. If you do your own meta-analytic study, I would recommend using the effect size measure that makes the most sense to the intended audience, can be explained most clearly, and can be calculated from the majority of worthy studies in the literature. For instance, as pointed out by Sun et al. (2018), when you have two possible outcomes in your sets of studies (e.g. success/failure, continue/discontinue, or survival/death), it would probably make most sense to use odds ratios or relative risks as the effect size. If most of the studies are addressing the relationship between variables, the researcher would probably use r, r^2, R, or R^2 for their effect size metric. If the researcher is interested in change or differences between groups (e.g. treatment versus control groups), they might choose to consider standardized mean differences such as Cohen's d.

7.1 The File Drawer Effect

While meta-analytic studies are very powerful and comprehensive, they do have at least one major drawback: the dreaded file drawer effect. The file drawer effect refers to the tendency for studies with nonsignificant results to be filed away and never published, particularly those with smaller sample sizes. Traditionally, primarily studies with significant results or those with very large samples are the ones that make it to publication. That is because nonsignificant results don't necessarily prove anything since the lack of significance may result from design flaws or lack of power. However, sometimes nonsignificant findings really do result from a lack of true differences or associations between our variables of interest and not from a shortcoming of the research. In those cases, the nonsignificant findings are very important. If we were able to include all studies meeting our criteria for inclusion into a meta-analytic study regardless of significance and publication status, we would potentially obtain different overall results. That is because the effect sizes from the studies with nonsignificant findings would decrease the overall, average effect size of the included studies. The resulting combined meta-analysis would potentially be a truer reflection of reality.

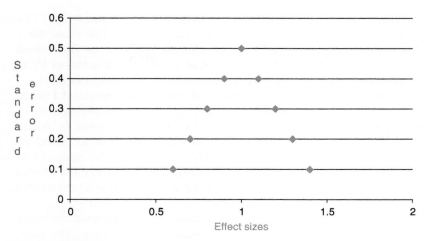

Figure 7.1 A funnel plot of effect sizes.

Sun et al. (2018) and Haidich (2010) both review strategies for testing for the presence of the file draw effect, which they refer to with the more technical term of "publication bias." For example, funnel plots allow the "meta-analyzer" to assess whether there appears to be a publication bias by plotting the effect size for each of a set of studies against their sample sizes. The *x*-axis (horizontal line) of the scatter plot is represented by the effect size, small to large, and the *y*-axis (vertical line) is represented by the standard error – since larger studies will have smaller standard errors, the standard error reflects the sample size. If there is no bias, there should be similar numbers of small studies representing both smaller and larger effect sizes. A publication bias would be indicated by the presence of more small studies in the distribution of larger effect sizes versus the smaller effect sizes. The premise is that smaller studies with negative results would have been less likely to be published than larger studies with negative results, while larger studies would have been published regardless of the outcome. When there is no bias, the scatter plot will have the appearance of an upside down, or inverted, funnel due to the symmetry of standard errors around the median effect size (see Figure 7.1).

7.2 Analyzing the Meta-Analytic Data

Once the studies have been selected and the effect size recorded or calculated for each study, the summary effect size is calculated by one of various possible methods. In general, these methods use a weighting procedure, applying weights to each study according to its sample size and, thus, variance in the scores. Studies with less variance in the scores are accorded more weight. The summary effect size is ultimately the average of the weighted individual effect sizes. Typically, the authors will also provide the variance of the overall estimated treatment effect so that the confidence interval of the overall effect size can be provided (i.e. the likely range in which the true effect size falls).

If you want to read the most straightforward, briefest description of a meta-analysis, see the online post called *Effect Size FAQs* (Ellis, 2010). In this easy-to-comprehend description, Dr. Ellis explains how to calculate the mean estimated effect size involving studies that report correlation coefficients. First, he weights each study's correlation coefficient (r) by multiplying the value of r by its respective sample size. He then adds the weighted coefficients and divides that sum by the total sample size (the sum of the samples sizes of all the studies combined).

As noted earlier in this chapter, the calculation of the mean weighted effect size is slightly more complicated when the individual effect sizes come from studies that vary with respect to the populations being considered. Again, with one population under consideration, the method will be referred to as a fixed effects procedure, while the procedure used to calculate the mean weighted effect size across multiple, varied populations is referred to as a random effects procedure.

Somewhat related to the issue of variations in populations being considered is the concept of statistical heterogeneity. For a good description of statistical heterogeneity and how to test and report it, see *How Can We Tackle Heterogeneity in Meta-Analysis?* (Nirphirake, 2018). Basically, in this context, heterogeneity refers to the degree to which true effect sizes vary across studies. Some of the variance will be related to sampling error, while some will result from actual differences in effect sizes across the studies selected. For a summary effect size to be meaningful, it is important for there to be minimal heterogeneity.

As discussed by Nirphirake (2018) and Sun et al. (2018), there are different methods for analyzing and reporting heterogeneity of the effect sizes. Cochran's Q statistic and I^2 are two such measures. The former tests for *lack* of heterogeneity, so a nonsignificant Q statistic means there is homogeneity, while a significant Q statistic means there is heterogeneity (which we do not want). Q is the sum of the weighted differences between the individual effect sizes and the pooled effect size (Stats Direct, 2005–2018a). The I^2 tests the degree of heterogeneity. In fact, the I^2 is really an extension of the Q statistic ($I^2 = 100\% \times \{Q - df\}/Q$ – where df is the degrees of freedom, or the number of studies minus 1) and it can be thought of as a ratio of the true variation across studies over the total variation (including chance variation due to sampling error).

If that does not make a lot of sense to you, you are not alone. If you were to conduct your own meta-analytic study it would become more concrete for you and make more sense. In the meantime, the important take away when you are judging someone else's meta-analysis is that you are looking for smaller degrees of heterogeneity in order for any summary effect size to have meaning. If $I^2 = 0$, then there is no heterogeneity (which would be a really good thing). According to Nirphirake (2018), I^2 of 0–25% indicates low heterogeneity (also good), while I^2 of 25–75% is moderate, with $I^2 > 75\%$ indicating high heterogeneity (which would not be so good in terms of the meaningfulness of the summary effect size).

A third method to evaluate heterogeneity of effect sizes as described by Nirphirake (2018) as well as Hoffman (2015) involves the use of forest plots to evaluate whether there is heterogeneity. A forest plot is a table or figure that illustrates each study's effect size and confidence interval relative to the summary effect size and its confidence interval. If the overall effect size can be seen in the graph to be included within the confidence intervals of all of the individual studies' effect sizes, the studies are deemed homogenous. If the overall effect size is not included within the confidence interval of all of the individual studies' effect sizes, then the studies are considered heterogeneous.

7.3 How to Read and Interpret a Paper Reporting a Meta-Analysis

Meta-analytic studies, if done well, will offer the reader a good review and technical summary of the literature addressing the topic being studied. Hopefully, the above discussion of meta-analysis will guide you in interpreting and judging the validity of a meta-analysis. The authors should present a detailed description of the criteria and methods they used to select studies, and some will even include a summary flow chart of their selection methods. There may be a separate table with characteristics of each study selected, including with respect to participant descriptors and study methodology. Finally, an item commonly included in many papers presenting meta-analytic studies is a nice, sometimes voluminous table of effect sizes across studies along with related metrics (e.g. variance, confidence intervals, etc.). At least one of the tables will commonly include the citation of each study (with author(s) and year published), and it is important to have access to sample size and characteristics of the research design. Tables 7.1 and 7.2 are examples of meta-analytic tables. Keep in mind these examples are fictional and do not depict any real events or actual people (okay, yes, I watch too much TV).

With regard to the criteria used to select studies for inclusion of a meta-analysis, some authors may refer to the Cochrane risk bias tool (Higgins & Altman, 2008). This is a tool composed of a set of criteria against which to judge studies and their designs. The tool was the result of a collaboration between a group of statisticians, epidemiologists, and others responsible for

Table 7.1 Sample layout of a meta-analytic table with standardized mean differences.

First author	Year	Patient *N*	Patient mean[a] (SD)	Control group *N*	Control mean[a] (SD)	Cohen's d	95% CI
Cagney	1975	50	19 (4)	50	40 (5)	−4.64	−5.39 to −3.89
Lacy	1978	30	22 (4.5)	28	38 (4)	−3.75	−4.61 to −2.90
*	*	*	*	*	*	*	*
*	*	*	*	*	*	*	*
Columbo	1973	40	21 (4.5)	40	39 (4.5)	−4.00	−4.76 to −3.24

a) Beck Depression Inventory score means and standard deviations.
* Signify that there are more lines of data not included.

Table 7.2 Sample layout of a meta-analytic table with *r*.

First author	Year	Sample	Blinding	SES	*n*	*r*[a]
Pierce	1980	Community	No	High	50	0.52
Winchester	1979	Institution	Yes	High	35	0.49
Father	1982	Institution	Yes	Mid	120	0.48
Klinger	1986	Institution	No	Mid	8	0.3
Houlihan	1981	Community	Yes	Low	100	0.47

a) Correlation between hours spent exercising per week and scores on a wellness measure.

evaluating research studies (see Higgins et al., 2011). The tool is really a systematic two-part process for assessing for risk of bias across six domains. These domains are: sequence generation and allocation concealment (selection bias and random assignment), blinding, incomplete outcome data (attrition and exclusion), selective outcome reporting, and "other issues" (any threat to validity, including fraudulent data) (Higgins & Altman, 2008). The first part of the analysis of bias includes describing characteristics of the study in detail as they relate to each of the six domains. The second part of the process involves assigning a value judgment as to the degree of risk for bias in each area (e.g. high, low, or unclear; see Higgins et al. (2011)).

When authors use this tool, they will discuss each study they include (or choose to exclude) in terms of the set of domains that address potential biases related to internal validity. When authors of a meta-analytic study provide a written description of their application of the Cochrane risk bias tool, they are offering transparency and detail that allow you to judge the quality of the studies that have been included. They are also demonstrating that they employed a clear and systematic review of the included studies.

7.4 Statistical Software Packages for Conducting Meta-Analyses

For those of you conducting your own meta-analytic study, you may want to find a statistical software package that can do the heavy lifting for you. If you do a search online, you will quickly see that there are many, many options for statistical software packages for conducting a meta-analysis. There are even online calculators that can be found with a simple online search.

If you are familiar with R, which is an open-source statistical environment, you will be happy to know that it can be used for meta-analyses. I would recommend the paper *A review of meta-analysis packages in R* (Polanin, Hennessy, & Tanner-Smith, 2017).

Sun et al. (2018) provide a description of *Comprehensive Meta-Analysis*, a commercial software package that can be purchased online. SAS® and Stata® also offer procedures for conducting meta-analyses. If you prefer SAS, you can actually search online and find open-source macros for using SAS with a variety of different meta-analytic techniques. Sheu and Suzuki provide examples of using SAS Proc Mixed for conducting meta-analyses (Sheu & Suzuki, 2001). If you are an SPSS® aficionado (or you just like using SPSS), you will probably find the paper *How to do a meta-analysis* (Field & Gillett, 2010) quite helpful.

7.5 Summary

Meta-analytic studies provide a useful tool for summarizing and analyzing findings across studies when those studies address a common question. There are a variety of techniques available for conducting and reporting meta-analyses, and these techniques will often be driven by the available data and metrics reported in the literature. These summary analyses allow us to consider whether research findings have been consistent across sites and samples. The ideal meta-analytic study will include well-designed studies, and the reports will provide detailed descriptions of the methods used to select and analyze those studies.

Section 4

**Reporting, Understanding, and Communicating
Research Findings**

8

Disseminating Your Research Findings

Once a study has been completed, it is time to share the results with the target audience. For many, the goal of research is publication in a scientific journal. Research can also be presented at conferences. In this chapter, I will review the basic mechanics of disseminating research findings in print and at conferences.

First, though, I want to point out an issue that many have grappled with in the sciences. That is, we typically only publish significant findings, though the nonsignificant findings may be equally important. This becomes most apparent when meta-analyses are performed to ascertain typical effect sizes in studies testing a particular hypothesis. Because nonsignificant findings are often not published, the published research will be skewed toward positive (i.e. significant) findings, and typical or average effect sizes will be biased. Remember, the dreaded file drawer effect (i.e. publication bias) I discussed in Chapter 7? To combat this problem, and to perhaps inform other researchers as they embark on a line of inquiry, it is important that we consider disseminating research findings even when they do not turn out as anticipated. While nonsignificant findings are not as exciting as significant ones, they are informative, nonetheless.

8.1 Preparing a Research Report

So, the data are collected and analyzed. It is now time to prepare a report. Reports can be written in different formats, and you should tailor your format according to the journal to which you wish to submit your work. When you have chosen the journal to which you wish to submit your paper, it is advisable to check the online instructions for authors. I would also recommend reading at least a few papers published in the journal (if you have not already) in order to get an idea of the expected format. Also, keep in mind that many journals have online submission programs.

Research Design and Analysis: A Primer for the Non-Statistician, First Edition. Leslie D. Rosenstein.
© 2019 John Wiley & Sons, Inc. Published 2019 by John Wiley & Sons, Inc.

In the following pages, I briefly review the format described in the *Publication Manual of the American Psychological Association* (2010) as many journals require APA style formatting. If the journal to which you are submitting your paper requires APA style formatting, it might help to borrow or purchase a copy of the manual for more detailed information. In a nutshell, though, you will need a title page, an abstract, an introduction, a methods section, a results section, a discussion section, a set of tables and figures (if any are needed), and a references section. The APA publication manual also specifies formatting, such as spacing, font size, margin widths, page numbering, and headers. The references also need to be formatted in a specific manner. Fortunately, if you use Word®, you are in luck because it offers an APA Publication style template.

After you have crafted an informative title and prepared your title page according to the APA format or the format required by the journal, you will prepare an abstract. Many journals have specific criteria for components and formatting of the abstract. Generally speaking, the abstract will be a brief summary of the study. There are often very specific word limits for the abstract. The abstract should contain a statement of the purpose of the study followed by a brief description of the participants/subjects and methods, a summary of the results, and a brief discussion of the meaning and importance of the findings moving forward. Table 8.1 provides you with the abstract of a sample study.

The next component of the paper will be the Introduction. The Introduction basically includes three major components. First is an explanation of the importance of the topic and why it should be studied. This is followed by a presentation of the literature review; that is, a summary and discussion of

Table 8.1 Abstract of a sample study.

Abstract
Objective: The purpose of this study was to evaluate the effectiveness of sleep hygiene and exercise education in improving mood and cognition among a diverse sample of patients in an outpatient neuropsychology clinic. **Method:** 136 participants recruited from an outpatient clinic received either care as usual, sleep education, exercise education, or combined sleep and exercise education in a single session and in writing. The focus of the sleep education was avoidance of excessive time in bed as well as sleep hygiene. Scores on a measure of mood and a measure of psychomotor speed and divided attention were obtained at baseline and at six months follow-up along with self-reports of exercise and sleep variables. **Results:** Groups receiving education about sleep and/or exercise showed improvement in mood and divided attention across the study period in comparison to the care as usual group ($p < 0.001$). At the end of the treatment period, groups receiving exercise education spent significantly more time exercising ($p < 0.001$). Those receiving sleep education spent less time in bed ($p < 0.001$), generally within the recommend range of 7–8½ hours per night. They were also less likely than the other groups to sleep with their TV or lights on ($p < 0.01$). **Conclusions:** Sleep hygiene and exercise education may be effective tools in changing behavior and improving mood and cognition. Intervention implications are discussed. **Key Words:** Sleep, Exercise, Cognitive, Depression, Neuropsychological.

relevant literature that has led to the specific study. Finally, the hypotheses are presented with a brief description of the method used to test them. See Table 8.2 for an Introduction section of a sample study.

Table 8.2 Introduction section of a sample study.

Sleep issues are often overlooked as primary contributors to depression and cognitive dysfunction in patients with medical and, more specifically, neurological conditions. Sleep issues may include: excessive time in bed, irregular sleep schedules, inadequate restorative sleep, and breathing-related sleep disorders. Additionally, patients' reports of poor sleep quality may be taken at face value and treated pharmacologically when, in fact, the patients are simply trying to sleep too much or have poor sleep hygiene (e.g. sleeping with the TV on). Oftentimes, when patients report difficulty falling asleep or maintaining sleep, they are actually trying to sleep more than is recommended or required, and their sleep onset and maintenance are actually normal. In these cases, adjusting their sleep schedule will be more effective than adding medications and will have fewer, if any, adverse side effects.

> Explanation of the importance of the topic

Excessive sleep is known to be harmful with respect to both physical and mental/cognitive health (Benito-León & Bermejo-Pareja, 2013; Liu, Wheaton, Chapman, & Croft, 2013; Sabanayagam & Shankar, 2010). Importantly, though, both short duration (<7 hours per day) and long duration (>9 hours per day) sleep have been linked to increased risk of cognitive impairment and/or dementia (Yaffe, Falvey, & Hoang, 2014). Excessive daytime sleepiness, not surprisingly, has also been linked to cognitive impairment in older adults even after controlling for related contributing factors affecting sleep (see Yaffe et al., 2014). Additionally, as expected, sleep-disordered breathing has been linked to poorer cognitive performance among older adults, particularly with respect to vigilance and executive functions (Yaffe et al., 2014). Sleep apnea has also been linked to depression (Hayley et al., 2015).

> Literature review

There is also currently a growing body of literature demonstrating the positive impact of physical activity on emotional health (see Craft and Perna (2004) for a review) and cognitive functioning (see Ratey and Loehr (2011) and Bherer, Erickson, and Liu-Ambrose (2013) for reviews; and Baker et al. (2010), Chapman et al. (2013), Lautenschlager et al. (2008), Vidoni et al. (2015)). Vidoni et al. (2015) found a positive impact of aerobic activity on measures of span of attention and working memory in a prospective, randomized controlled trial. In addition, though, in this community-based pilot study, they found a dose–response of exercise duration on some of their additional cognitive measures. In young adults, the relationship between habitual exercise and cognition may be mediated by cerebral blood-flow regulation (Guiney, Lucas, Cotter, & Machado, 2015).

In spite of the positive findings regarding exercise, cognition, and mood, exercise is infrequently a focus of treatment of patients with depression and/or neurological compromise. When exercise is recommended, patients are generally not provided concrete direction in terms of frequency, duration, intensity, and type. Instead, patients may be placed on medications (e.g. antidepressants) that may have significant side effects and are not always adequately effective. Assisting patients in changing lifetime behaviors and improving

(Continued)

Table 8.2 (Continued)

their level of physical activity offers a tool that can be carried forward to manage depression and some cognitive dysfunction in patients with comorbid neurological conditions.

As more studies demonstrate the relationship between sleep variables and emotional and cognitive functioning and between exercise and emotional and cognitive functioning among patients with neurological conditions, it is hoped that practitioners will become more alert to the need to attend to sleep and exercise as a first-line of intervention. Demonstrating these relationships among a patient population additionally supports the need for clinical trials in which patients can be assigned to sleep- and exercise-intervention versus standard-treatment groups.

There is already some emerging evidence among clinical groups supporting exercise programs as an effective intervention. For example, David et al. (2015) recently showed that exercise may improve cognition in Parkinson's disease, with patients in two exercise interventions showing improvements in their measures of span of attention and working memory. In a meta-analysis, Kalron and Zeilig (2015) found 12 controlled clinical trials investigating the impact of exercise-intervention programs on cognitive functioning in groups of patients with Parkinson's disease (PD; $n = 3$), stroke ($n = 1$), and multiple sclerosis (MS; $n = 8$). The findings were mixed, with nine of the 12 studies finding at least some significant positive impact of exercise training (including the stroke group, all three PD groups, and five of the MS groups).

The current study was intended to further assess the association of sleep and exercise with cognitive and emotional functioning in an outpatient neuropsychology clinic sample with a range of diagnoses. Four groups of patients either given care as usual, exercise instruction, sleep education, or exercise and sleep instruction were compared on measures of cognition and mood. It was predicted that the three education groups would show improved mood and cognition after a six-month period, with the most improvement occurring among those receiving both exercise and sleep instruction. Given the broad range of conditions and severity in the current clinical sample, effects of sleep and exercise would need to be particularly strong to overcome variance attributed to other condition-related factrs.

Following the Introduction comes the Methods section. The Methods section is generally divided into subsections. The first subsection includes a detailed description of the sample(s) (Participants). This includes sample size(s), major demographics of the sample, and inclusion and exclusion criteria for participants. The Participants section should include a description of how the participants or subjects were recruited and then assigned to groups. Finally, this section should include any power analyses that were conducted to determine the ideal sample sizes. See Table 8.3 for an example of a Methods section.

The second subsection of the Methods section includes a description of any materials used in the study (the Materials). This subsection includes a listing and description of any measure or instruments used. If a relatively new instrument or measure is used, this subsection should also include information about the reliability and validity of the tool. If a new questionnaire was developed for

Table 8.3 Sample methods section.

Method

Participants

   136 consecutive patients seen in an outpatient neuropsychology clinic in a large medical center were recruited and randomly assigned to one of four groups (care as usual, exercise instruction, sleep instruction, and combined sleep and exercise instruction) with an equal number of participants per group. Sample size was planned based on the result of a power analysis using G*Power (Faul, Erdfelder, Lang, & Buchner, 2007) with an estimated effect size of 0.5, power set at 0.8, and alpha of 0.05. Exclusionary criteria included being under age 18, having delirium, suffering a stroke within the previous 30 days, or undergoing neurosurgical intervention within the previous 30 days. Participants ranged in age from 18 to 89 years with a mean age of 53.3 years (SD = 15.51). Education ranged from 5 to 22 years with an average of 14.3 years (SD = 2.94). There were 69 males and 67 females. There were no statistically significant differences between the groups in terms of education ($F_{[3,132]} = 1.2$, $p > .10$), age ($F_{[3,132]} = 1.3$, $p > .10$), or gender ($\chi^2 = 0.03$, $p > .10$). Diagnoses of the patients are listed in Table *X*.

Materials and Procedure

   Participants were recruited following their clinic visit by a research assistant blind to the patient's diagnosis or test results. Participants were not informed of the predicted findings of the research. A second research assistant recorded participants' demographic information and their scores on two test measures. These were the Beck Depression Inventory II (Beck, Steer, & Brown, 1996) total raw score and the Trail Making Test, Part B (Reitan, 1958) time in seconds.

   The Beck Depression Inventory was used as a summary measure of self-reported depression symptoms. The Trail Making Test B was used as a measure of visual scanning and divided attention, or set shifting ability (Trails B). In cases in which either trial of the Trail Making Test was discontinued due to excessive time and difficulty, the scores (total time in seconds) were prorated using the methods outlined in Heaton, Miller, Taylor, and Grant (2004). Specifically, the time score in seconds was divided by the number of circles completed and then multiplied by 25.

   Following the consenting process, participants assigned to the sleep education group were given education and recommendations. See Appendix *X* for the printed informational sheet that was provided with the oral presentation. The education and materials explained the reasoning behind the recommendation to sleep with the TV and lights off, as well as information regarding the adverse consequences of spending more than 8½ hours in bed per night. Additional information about sleep hygiene was included as well.

   The participants assigned to the exercise education group were given education and recommendations, both orally and in writing (See Appendix *XX*). In addition to education about the benefits of exercise, they were advised to begin exercising five days per week. They were instructed to exercise 45 minutes per day, gradually working up from 10 minutes per day if they were not already exercising regularly. They were instructed to follow their physician's recommendations regarding any restrictions on the type of exercise they chose.

(Continued)

Table 8.3 (Continued)

The participants in the combined exercise and sleep hygiene group were provided with both sets of materials along with orally presented education and instructions. The participants in the care as usual group did not receive any materials or extra education.

Six months after each participant had been recruited for the study, they returned for a prescheduled appointment. At this follow-up appointment, they met with the first research assistant and completed a questionnaire along with the Beck Depression Inventory-II. The research assistant also administered the Trail Making Test. The questionnaire included self-report information about typical time spent in bed, whether the TV and/or lights were kept on during the night, days exercised per week, average duration of each exercise session, and type of exercise. All patients completed the Trail Making Test first, followed by the Beck Depression Inventory-II, and then the questionnaire in order to minimize reactivity and potential bias.

Statistical Analyses

Two 4×2 repeated measures analysis of variance (ANOVA-RM) were conducted. The first analysis assessed differences between the four groups in change over time in Beck Depression Inventory scores. The second analysis assessed differences between the four groups in their change over time on Trails B performance. The data could also have been analyzed using Mixed Linear Model analyses, but with the fully balanced design, no missing data, and no major assumption violations, the results would have been equivalent. In addition to the primary analyses, the groups were compared using a single-factor analysis of variance (ANOVA) to assess for any differences in typical time in bed per night and typical weekly exercise (days per week multiplied by duration per session). All four analyses of variance were followed with comparisons between each pairing of the four groups at baseline and also at the end of the six-month treatment period using a Bonferroni correction; this conservative measure was chosen given the elevated risk of Type I errors with such a large number of comparisons. Finally, two Chi-Square analyses were calculated to assess for any differences in the number of participants in each group reporting that they sleep with the TV or lights on.

the study, a description of how the instrument was developed and tested should also be included.

The Procedures subsection includes a brief description of the design as well as details about how the study was conducted. Depending on the complexity of the design, some papers include a separate subsection (Design) before the Procedures section, and some journals may even require that. In general, the Procedures section should provide enough detail that someone replicating the study will know exactly how the study was conducted.

Some journals require a description or listing of planned statistical analyses to be included as a subsection (Statistical Analyses) at the end of the Methods section. This especially makes sense if the analyses will be complex. A description of less-complex statistical analyses can also simply be incorporated into the Results section.

Next comes the Results. The Results section includes the results of statistical analyses. This may include a presentation of descriptive statistics (e.g. group means, standard deviations, and sample sizes), which can be presented in a table. This is followed by a description and the result of each statistical analysis conducted to test the hypotheses of the study. The results of analyses typically include a verbal characterization of the analysis as well as the actual numerical result of the analysis. The latter can be provided parenthetically, and should include the statistic, the p value, and the effect size. Information about confidence intervals can also be included; in fact, this is recommended in the APA (2010) publication manual. Table 8.4 includes the major components of a Results section from a sample paper.

The last major section of a study paper is the Discussion section. The Discussion is generally composed of four elements. The first component is a brief summary of the findings reported in the Results section. Here, though, the numerical results of the statistics are not repeated; rather, the general findings

Table 8.4 Sample fictional results section.

***************These results are completely fictional***************

In the first repeated measures analysis of variance, there was a statistically significant Time × Group interaction ($F_{[3, 132]} = 88.91, p < 0.001$) indicating differences across the groups in the degree of change in Beck Depression Inventory-II scores from baseline to post-intervention. As can be seen in Figure Y, the three intervention groups showed decreasing depression scores from baseline to post-intervention, but the care as usual group did not. Moreover, as can be seen in Table XX, there were significant differences between all three treatment groups in comparison to the care as usual group at the end of the six-month period. The combined sleep and exercise education group had significantly lower depression scores compared to all three groups. The sleep education and the exercise education groups each had lower depression scores than the care as usual group. There were no significant differences between the groups at baseline. The same pattern was found for Trails B scores as can be seen in Table XXX. The Time × Group interaction was also significant ($F_{[3, 132]} = 93.52, p < 0.001$) again indicating differences in degree of improvement from baseline to post-intervention across groups (see Figure YY).

At the end of the six-month period, the four groups differed significantly in their hours spent in bed per night ($F_{[3132]} = 65.41, p < 0.001, \eta^2 = 0.74$). As can be seen in Table XXXX, the combined sleep and exercise education group and the sleep education group both reported less time in bed per night than the other two groups, and they did not differ from one another. As reported in the table, the groups receiving sleep education reported sleeping within the recommended guidelines, on average, of 7–8½ hours per night. The four groups also differed significantly in the number of hours they reported exercising per week ($F_{[3132]} = 73.13, p < 0.001, \eta^2 = 0.79$). Here, as can be seen in Table XXXXX, the combined group and the exercise education group both reported exercising significantly more hours per week than the other two groups; they reported exercising, on average, within the recommended guidelines of 2½–5 hours per week.

The four groups also differed in the frequency with which they reported sleeping with the TV on ($X^2 = 44.34, p < 0.001$) and the lights on ($X^2 = 11.21, p = 0.01$). As can be seen in Tables XXXXXX and XXXXXXX, the sleep education and the combined groups were less likely to sleep with either TV or lights on than the other two groups.

---

are verbalized. The second component of the Discussion is a presentation of any limitations of the study. This may include issues with generalizability and lack of ability to conclude causality. Problems with attrition and recruitment, if any, can be included in the discussion of generalizability. Lack of control group and related issues should be included in the discussion of causality, if relevant. The next component of the Discussion is a summary of the importance of the findings, and how they may apply going forward. For instance, a successful clinical intervention study may influence choice of intervention for patients with the condition under study. Finally, the Discussion section should include suggestions for future research. See Table 8.5 for a sample Discussion.

Table 8.5 Discussion section of a sample research report.

<table>
<tr><td align="center">**Discussion**</td></tr>
</table>

Summary

In this study, both sleep education and exercise education were associated with improved performance on a measure of divided attention. The two types of education were also associated with a decrease in self-reported depression symptom severity. The combined education seemed to have the biggest effect on both depression scores and divided attention. Analyses also indicated that individuals receiving sleep education were less likely than others to sleep with the TV or lights on, and they spent less time in bed than the others. The individuals receiving exercise education also reported spending more time exercising per week. These latter findings suggest that the education interventions had an effect on self-reported behavior in addition to mood and divided attention.

Future Directions

The current study supports the notion that too much sleep is associated with negative outcome, and exercise is associated with less depression. Future studies might explore predicting who is less likely to exercise, to have poor sleep hygiene, or to spend too much time in bed. Another potential direction for research could be to determine whether the individualized education programs here could also be effective in a large group format or with mass communication, such as through Public Service Announcements, brochures, etc.

Additional studies may also involve educating families and providers on the importance of exercise as a first line of treatment in depression, and on the importance and relative safety of spending less time in bed (to a point). Providers' initial reliance on pharmacological management of depression and insomnia versus education and support could then be assessed pre- and post-education. Patients' participation in exercise, reports of general wellbeing, and reliance on pharmacological management of depression and/or insomnia could also be measured pre- and post-education.

Implications

A link between exercise and mood has been shown repeatedly. Given that exercise is generally not harmful, assuming the exercises are appropriate for a given patient and his/her injuries and other conditions, it follows that regular exercise should be considered a regular first-line intervention in the treatment of depression among patients with a variety of neurological conditions.

Table 8.5 (Continued)

The current study along with previous studies, including those discussed earlier, also suggests that time in bed should be discussed and addressed with patients. That is, patients may be educated regarding the ideal time in bed, and cautioned against spending too much time in bed. When they are sleeping more than nine hours per night, this may be an indication of level of impairment, but consideration should also be given to the possibility that the increased sleep is adversely impacting the level of functioning, at least cognitively and emotionally. In clinical practice, there may be a tendency at times to help patients increase their sleep when they complain of insomnia, without awareness of the time actually being spent in bed and its ramifications.

The irony is twofold. First, patients become concerned that they have insomnia. They may then turn to over-the-counter sleep aids, many of which have side effects that may worsen mood or day-time energy. Some of these medications also may contribute to cognitive issues, such as those containing anticholinergic compounds. Excessive sleep has been linked to increased mortality, and use of pharmacological sleep aids has been linked to increased mortality as well. In a study of over one million men and women aged 30–102, Kripke, Garfinkel, Wingard, Klauber, and Marler (2002) found a 15% increased mortality hazard in individuals exceeding 8.5 hours sleep or sleeping less than 3.5 or 4.5 hours. And, while reporting insomnia was not associated with excess mortality hazard, using prescribed sleeping pills was associated with significantly increased mortality after controlling for sleep duration and insomnia reports. While the authors concluded that causality is unproven in terms of the increased risks, they did note that patients can be reassured that short sleep and insomnia do not increase mortality.

Conclusion

The current study highlights the need for focus on sleep duration and exercise with respect to mood and cognition among neurologically compromised individuals. Further, relatively safe and inexpensive interventions involving education and follow-up may be effective in making significant changes in sleep and exercise behavior.

Also included in a paper formatted for research will be a References section. This will be followed by any tables and figures. Generally, when a paper is submitted for publication, the author(s) refer to each table in their text (e.g. "See Table X" as you may have noted in the sample Results section), but the actual tables are included, each on a separate page, following the References section. Any figures are provided in the same manner, following the tables. Finally, the journal to which you are submitting may have very specific guidelines for formatting of the References section, tables, and figures.

8.2 Presenting Your Findings at a Conference

Findings of studies are also often presented at conferences. This enables dissemination of research findings during the delay that will inevitably occur between completion of data collection and publication in a journal. The delay is the result of the write-up time, the submission and review process, any revisions, and then the publication lag (which has actually been greatly reduced with the advent of online publishing).

Presentation at conferences typically takes one of two forms, either a poster presentation or a paper presentation. A paper presentation is an actual presentation to an audience that may occur as part of a symposium or topical session. A poster is a pictorial presentation of a study, with text and graphics mounted or printed on a medium the size of a typical poster.

Posters are typically presented in an exhibit hall. There are often dozens or more posters displayed in the exhibit hall during a poster session. There may be more than one poster session at a conference, and each poster session may include clusters of posters on related topics. Presenters typically stand next to their poster in order to answer questions, and some even hand out a paper copy of the poster or a brief summary of the study. It is a great opportunity for sharing information and networking.

Posters are selected during a submission process. The submission guidelines and deadline are typically posted on a professional society's website along with meeting/conference information. Submissions are usually made in the form of an abstract, and that abstract is often published in the conference proceedings if the poster is accepted for presentation. Once a study has been accepted for presentation, the actual poster is created. The format and size of a poster is mandated by the conference organizers.

8.3 Summary

Once a study has been completed, the findings are typically shared with the scientific community, the public at large, or both. I have described two methods for sharing written reports with the scientific community. These methods are formal written reports to be submitted to journals, and preliminary reports that can be submitted in a poster format at academic meetings and conferences. As noted, research can also be shared with the scientific community through formal presentations at conferences. Sometimes, our research, if of particular interest to the general public, may be reported through press releases, social media, news reports, media interviews, and newsletters. It is in these instances in which our presentations must be clearly conveyed in a manner that can be understood by all, and without the potential for misuse. I will address this concept further in Chapter 9.

9

Concluding Remarks

The goal of writing this primer was to provide a broad overview of research techniques, from ethical considerations to the design process, analysis, interpretation, write-up, and dissemination of findings. The purpose of this overview was to provide assistance to researchers without a strong background or interest in statistics who may wish to design their own studies. A second, equally important purpose was to assist those who wish to become good consumers of research. Those who strive to be good consumers of research may include clinicians, educators, policy makers, and journalists whose reporting is focused on the sciences.

In the design and analysis chapters, I intentionally reviewed some of the more basic research designs that may be considered by researchers in the clinical and behavioral sciences and in education. Some research fields, particularly in the biomedical sciences, agriculture, and economics, will have more complex designs with more variables to consider.

If you are new to research and are interested in conducting your own studies, I strongly recommend consulting with a statistical consultant before you even finalize your design and begin collecting data. In case you are interested in furthering your knowledge and understanding of statistics, I have included a few resources in Appendix C.

9.1 Why Is It Important to Understand Research Design and Analysis as a Consumer?

For clinicians and educators, a good understanding of research design and analysis will aid in evaluating research reports. This is important especially if you plan to alter or determine options and interventions for patients or students based on the results of research studies. If a study is flawed, you could make the wrong choices for your charges. Sometimes, this will have benign

Research Design and Analysis: A Primer for the Non-Statistician, First Edition. Leslie D. Rosenstein.
© 2019 John Wiley & Sons, Inc. Published 2019 by John Wiley & Sons, Inc.

consequences, and sometimes the consequences can be serious or at least result in unnecessary expenses in terms of fiscal costs and time.

One particular issue we face as clinician consumers of research is the translation of research findings to clinical practice with an understanding of group statistics versus individual predictions. With the exception of case studies, research studies tell us about groups, not individuals. That is, when we conduct statistical analyses, we are investigating differences between samples or relationships between variables for a sample. How well research findings generalize to other groups or outside of the study sample will depend on how well the sample represents the population to which we want to generalize.

Moreover, how well the research findings generalize from groups to individuals will depend on the spread of the distributions and normal variation. That is, the degree to which a clinician can make predictions for an individual patient or student based on statistical analyses using a sample will depend on how much variance there is in the sample and how much variance there is in the population. If there is normally quite a bit of variation (often referred to as *normal variation*) in the population, then it will be quite difficult to predict how an individual will respond to an intervention, for example, based on the findings of a research study that applies statistical analyses to groups of people. That is because it may be difficult to determine where on the normal distribution the current individual falls with respect to the predictors and the outcome being pursued and predicted. However, if it is possible to increase the specificity of a study sample by defining and dividing groups based on multiple characteristics (including so-called extraneous variables), it will be possible to increase the accuracy of a prediction involving a single individual.

For example, if we know from research that patients with a particular diagnosis, in a particular age range, who are of a specific gender, and with or without symptoms A, B, and C respond favorably to Treatment 1 versus Treatment 2, then we can use those specific characteristics to make predictions about a single, individual patient. If the study simply considers patients with the particular diagnosis but without defining and assessing the impact of the additional characteristics (variables), then we will not be able to be as confident in making predictions for our particular patient. The better our individual patient matches the sample or subgroup studied in the research, the better we will be able to generalize from the study to our patient. This is because the more tightly defined research has reduced the "noise" or extra variance unrelated to the actual treatment and the characteristics of the individual.

Unfortunately, clinicians sometimes make diagnoses, for example, based on findings from research addressing the presence or absence of a particular set of symptoms or test results without consideration of normal variation. That is, the clinician may rely on findings from previous research demonstrating the presence or absence of a set of symptoms in a given condition, but without considering that the presence of those symptoms is not uniform across each and every participant in the research. For example, the finding that many patients with a specific form of dementia tend to exhibit a particular pattern of results across measures means just that: there is a *tendency* for that pattern to occur in the condition, but not all people with that condition will exhibit that pattern. Additionally, not all people with that pattern of test performances will necessarily have the pathology associated with the condition; that is, there may be other conditions or factors that contribute to the pattern of findings. The research has just shown us that there is a link between a pattern of test results and a disease, but one cannot conclude with perfect confidence whether an *individual* has that condition based on the pattern of test results.

As clinicians, we must continue to be mindful of the distinction between group statistics and individual predictions. The group statistics improve our predictions about the individual, but the degree of precision in making an individual prediction will depend on the amount of variability within the population or even within groups with the condition of interest. We must also be mindful of the need to distinguish between associations between two variable or events demonstrated through research, and actual proven causality demonstrated through carefully controlled experimentation.

For reporters, a good understanding of research design and analysis is important in evaluating and reporting the results of research in a way that is meaningful and not unintentionally misleading to readers and viewers. Erroneous reporting or dissemination of erroneous interpretations of research can lead to great harm to both individuals and the population as a whole. Most often, the erroneous interpretations are as innocent as the overstatement of findings. However, there are occasionally blatantly false claims and/or falsified data reported by researchers, and good investigation and review by reporters can be critical.

One of the starkest illustrations of the harm that can be perpetrated by the recurrent reporting and sharing of false research claims may be the errant connection made between the measles, mumps, and rubella (MMR) vaccine and autism (see Belluz (2018), Flaherty (2011), and The College of Physicians of Philadelphia (2018), for discussions of the issues). The small, multi-case study that led to the erroneous claims of a link between the MMR vaccine and autism was flawed in its design, but also replete with apparent falsification of data. Moreover, at least one of the authors reportedly had financial conflicts of interest, making his recommendations for alternative single-disease vaccines even more dubious.

What is particularly notable is that the authors did state in their original paper (though not in later videos) that a causal relationship could not be established between the MMR vaccine and autism. Astonishingly, many individuals in the public as well as organizations have continued to perpetuate the misnomer that the MMR vaccine causes autism in spite of this caution and even following the later retraction of the original report from the journal in which it had been published (see Editors of The Lancet (2010)). The outcries against the MMR vaccine have continued in spite of countless studies being conducted that have ultimately debunked the claims that vaccines are related to autism (see Taylor, Swerdfeger, and Eslick (2014) for a meta-analytic review of the research).

At the same time, whether causal or not, vaccination rates have declined, and the mumps and measles are on the rise. This is reviewed by Majumder, Cohn, Mekura, Huston, and Brownstein (2015), who discuss the association between vaccination rates and a measles outbreak. The Centers for Disease Control and Prevention (2018) also provides a review of mumps trends. Both measles and the mumps can have very serious outcomes (CDC, 2018) and, in the case of measles in particular, even death (CDC, 2017b).

Unfortunately, the original research report and continued coverage of the overstated findings about a vaccine–autism connection have had a lasting, negative impact. Once overstated or erroneous research findings are reported and repeated, it is very difficult to reverse the harm that has been done.

Another area in which the overstating and misinterpretation of research findings has plagued the public revolves around the concept of *Chronic Traumatic Encephalopathy* (CTE) and its etiology and possible complications. In truth, the study of CTE is very early in its development, and while there is enough evidence to suggest that there should be continued investigation of the identified brain pathology, claims about its cause or its consequences cannot be made with confidence as of the writing of this primer. In spite of this, the media as well as researchers and clinicians have repeatedly promulgated an impression that CTE is definitely caused by repeated concussions or subconcussive blows, and that the end result is of a serious, debilitating, and unremitting condition. Inclusive, there have been claims that CTE in NFL football players is causally linked to suicide, and that CTE is prevalent among NFL players. In fact, research has refuted the presence of an increased incidence of suicide rates among NFL players (for example, see Lehman, Hein, and Gersic (2016)).

Brooke de Lench (2017) does a very nice job of characterizing the current state of the situation related to the lopsided dissemination of research findings with regard to CTE and the unwarranted degree of fear generated by the hype and sensationalizing. In an editorial in the *British Journal of Sports Medicine*, Kuhn, Yengo-Kahn, Kerr, and Zuckerman (2016) discuss the phenomena that have likely contributed to the misrepresentation of the current state of the science of CTE. In particular, they review the Semmerlweis effect, in which new evidence is downplayed, if not rejected due to those findings contradicting established beliefs and biases. This leads to what they refer to as the availability cascade, causing propagation of one-sided biased perceptions. They go on to admonish researchers and the media to be aware of our biases, and for researchers to disseminate our science to the media and to even encourage media attendance at academic conferences and meetings. They also encourage researchers to engage in writing letters and responses to journal editors and to engage in cordial discourse vis-à-vis back-and-forth debates at academic conferences.

A third area of medical science that has been covered extensively by the media and consumed by both the public and clinicians is that of hormone replacement therapy. In 2002, there was an initial report of a large-scale study of hormone replacement therapy in older, postmenopausal women called the Women's Health Initiative (WHI) (2002). Due to its large size and thoroughness, the findings in that publication were immediately reported extensively, and those reports seemed to take on particular urgency due to the Data Safety Monitoring Board initially discontinuing one arm of the trial at that point due to concerns about elevated risks of heart disease, cerebrovascular conditions, and breast cancer among a group of the older women. At the same time, notably, there were also findings of decreased risks of colorectal cancer, osteoporosis, and hip fractures among the same women. There were additionally findings of decreased risk of breast cancer in another group of women receiving a slightly different hormone replacement therapy regimen in the same initiative. These latter findings have not been reported nearly as extensively as the negative findings.

Based, in part, on summary reports of the initial WHI findings and the decision to discontinue part of the trial, many women and their clinicians have opted to stop or to avoid hormone replacement therapy (Lobo, 2013). However, a close inspection of the original study report reveals certain idiosyncrasies of the data. In particular, the participants in the treatment arm that was discontinued were older, postmenopausal women when they started their hormone replacement therapy with a median age of 63; this is reportedly not consistent with recommendations for starting hormone replacement at an earlier age closer to onset of menopause for its positive benefits (see Klaiber, Vogel, and Rako (2005) for a discussion of this issue). Klaiber et al. (2005) also point out issues with the combination estrogen–progestin that was used in the discontinued arm of the study, as well as the fact that the treatment was administered continuously rather than cyclically. More specifically, they note that previous research studies had found benefits of hormone replacement in which either estrogen was administered unopposed, or the progestin portion of the estrogen–progestin combination was administered only 10–14 days in each 28 day time period versus every day.

Another important consideration that has contributed to the misuse and overinterpretation of the WHI study is the manner in which the magnitudes of the findings were initially reported. The data were presented with repeated coverage of the study as indicating a substantial increase in breast cancer, heart disease, and stroke among women treated with hormone replacement therapy. However, the initial report had addressed the relative risk and not the absolute risk of those conditions, a difference not understood by most people without a strong background in statistics or epidemiology. Misunderstanding of these metrics has contributed directly to the overinterpretation and misapplication of the results.

These concepts and the resulting misreporting of the findings are discussed with very clear explanations and illustrations by Dr. James Clark (2006). He illustrates through one example, in particular, the difference between the relative and absolute risks of hormone replacement therapy on heart disease in this specific population as presented in the data. It had been reported that there was a 29% increase in heart disease in the combined estrogen plus progestin group in comparison to the placebo group. This increased risk amounted to a 0.07% increase in actual cases of disease. That is, 30 women for every 10 000 who were in the placebo group (no hormone replacement) had heart disease, and 37 out of every 10 000 women treated with estrogen plus progestin had heart disease (there were actually a total of 8 506 women treated with the hormone combination, and 8 102 who did not receive hormone replacement, but these numbers are transformed to the 10 000 base for ease of explanation). A 0.07% increase in cases is much less alarming than the impression that there was a 29% increase in risk.

To put that into context, one also should look at the impact of the combination hormone replacement therapy on the risk of hip fractures that had been demonstrated in the same study. In the same data set, nearly 8 of every 10 000 women who did not receive hormone replacement therapy suffered a hip fracture, while 5 of every 10 000 women treated with the hormone combination suffered a hip fracture (see Rossouw et al., 2002). There was a larger difference when considering all fractures combined (97 versus 76 per 10 000).

Giving consideration to the risk (or decreased risk in this case) of hip (and other) fractures is no less important than considering the risk of heart disease or breast cancer; that is, hip fractures are potentially quite harmful as well. For example, hip fractures are known to have serious if not grave consequences in the elderly. In a prospective study carefully controlling for the effects of pre-existing health status and cognitive functioning, Dr. Erin LeBlanc et al. (2011) found a significantly increased risk of mortality within the first year (even months) following a hip fracture in women over 65 years of age.

With popularly held beliefs about hormone replacement therapy based on the initial WHI reports and coverage, it may be surprising to learn about the more recent reports about *decreased* risk of breast cancer found in the same WHI research (as noted earlier). When considering women with prior hysterectomy enrolled in the WHI, treatment with estrogen replacement was actually associated with a decreased risk of breast cancer and no change in the risk of heart disease or stroke in that group of women (LaCroix et al., 2011).

Unfortunately, relative to the initial coverage of the 2002 report and the early discontinuation of more than one arm of the hormone replacement trial, reports that later uncovered and pointed out the flaws of the initial report as well as several benefits of hormone replacement have received very little coverage by comparison. Because of this, who knows how many lives have been adversely impacted due to decisions that have been made based upon erroneous interpretations of the WHI findings. Moreover, the coverage of the discontinuation of the study has potentially impacted pharmaceutical companies in the private sector with respect to their ability to conduct further investigations of hormone treatments in spite of much evidence of the potential benefits. For example, as suggested by Vicki Brower (2003), it may be difficult to recruit participants for large-scale studies given the prior bad publicity.

The issues of the MMR vaccine and autism, the causes and consequences of CTE, and the risks versus benefits of hormone replacement therapy are just three modern examples in which research reporting has had adverse consequences on individuals and society. In each of these examples, initial reports were misleading or overstated, and then there was a perpetuation of claims via researchers, clinicians, and the media. In most cases, the perpetuation and misuse of the data could potentially have been prevented or minimized with close inspection and critical review of the original reports. Generally, such critical review requires some knowledge and comprehension of research design and interpretation.

9.2 Research Ethics and Responsibilities of Journalists

In the introduction to this primer, I briefly reviewed journalists' ethical responsibilities with respect to science reporting. As stated, those of you who work as professional journalists are aware of your great responsibility in reporting research findings to the public. Journalists have a special role in providing information to the public while avoiding harm. This is eloquently outlined in the *Professional Journalists' Code of Ethics* (Society of Professional Journalists, 2014).

The *Professional Journalists' Code of Ethics* mandates that journalists are responsible for the accuracy of their reporting, including verifying the information before it is released. Again, as noted by Carrie Figdor (2017), journalists face difficulties in their role of reporting and providing information that is accurate when the material is the product of scientific endeavors. As illustrated in the cases of the autism–vaccine connection, CTE, and hormone replacement therapy, journalists cannot always rely on the authors of scientific reports to provide accurate and valid information. This issue is especially critical now, with the availability of mass communication tools and social media. Non-peer-reviewed research reports are more readily available to the masses, and journalists cannot necessarily count on peer-reviewed journals to publish only sound research.

It is imperative that journalists carefully review and understand the research designs presented along with the results and conclusions of studies on which they report. As just one example of the need for understanding, journalists and others must take care to not translate a conclusion of an association between two events or variables as a claim of causality. Some factors that can contribute to incorrect interpretation of an association as a causal relationship are third (extraneous) variables, history, and maturation. These factors impact internal validity of a research study's design. That is, these types of issues impact the conclusions that can be drawn about the relationship between two variables (see Chapter 2 for more discussion). In Tables 2.1 and 2.2, I provided several examples of hypothetical studies in which conclusions about causality can be drawn erroneously along with more accurate descriptions of the findings.

Oftentimes, the correct language addressing the presence of an association as opposed to a causal relationship is included in a research publication. However, it is incumbent on the journalist and other consumers to read and understand such language. Otherwise, there is a real and great risk that the public will be misinformed and harmed as a result.

Journalists and other consumers of research should also be wary of data that have been misanalysed, or statistics that have been misinterpreted. Related, as illustrated in the discussion of the Women's Health Initiative (2002), the metrics that are used should be clearly understood in order to be accurately portrayed in further discussions. Moreover, the degree of importance of findings should be critically considered in terms of effect size and clinical significance (discussed in Chapter 4). Remember that a data analysis may reveal a statistically significant finding, but this does not tell the whole picture in terms of the magnitude of the finding. A statistically significant finding indicates that there is a reliable or probable relationship or difference (depending on the analysis performed). It does not tell you how *big* that relationship or difference is, or whether it has any impact of importance.

To summarize, journalists and others in the media hold positions of great influence due to the nature of their jobs and the ease with which news is accessed by the public. This influence may be tenfold for individuals with celebrity status who take on the mantle of advocating for a specific group or condition and cite the research community. When research findings are misinterpreted or overstated, an avalanche of misinformation can ensue. Large swaths of people may make potentially harmful choices on the basis of their erroneous interpretations of research findings. By attaining education about and understanding of research methods and interpretation, those in positions of influence can prevent these types of situations. By understanding issues surrounding causality versus association, magnitude of effects versus statistical significance, and the impact of extraneous, third variables, for a few examples, journalists and celebrities can deliver information to consumers in a manner that improves knowledge and engagement in educated choices while avoiding potentiation of hysteria and the making of uninformed decisions.

9.3 Responsibilities of Researchers

I do not want to come across as though I think the media (or Hollywood) are to blame for all of the problems caused by misinterpretation of research or perpetuation of bad information. In fact, researchers carry the brunt of the responsibility. It is our responsibility to be extremely clear in conveying our research findings and not overstating claims.

To be fair, in fact, researchers are usually at the center of the quagmire. There is such a drive to publish or perish in order to get promotions, that publication and dissemination of research findings becomes the end game. The drive to publish and to make a name for oneself sometimes takes precedence over all else. As scientists and researchers, we have a great responsibility to ensure that we put the subject/target of our research first, and our own personal ambitions last.

As noted by Kuhn et al. (2016) and discussed earlier in this chapter, besides not overstating our findings, we need to be active in responding to and debating our fellow scientists even when we are concerned that our arguments will not be popular. We should engage in sound, scientific inquiry. It would also behoove us all to encourage and invite our colleagues in the media and even those with a platform due to their celebrity status to attend or be part of our scientific conferences. It is easy to point fingers and shake heads, but it is much more productive if we work together to educate and disseminate research findings accurately and responsibly.

In all, researchers should think carefully about the research we choose to conduct, questioning the purpose and potential benefit. When we do conduct research, we need to be clear in conveying our findings *and* their meaning. We need to inform our audience of any weaknesses or shortcomings of our research as well as the limitations of the findings. We must be aware, too, that our audience is not simply composed of other trained researchers who comprehend our lingo and jargon. In other words, we cannot flippantly state, for instance, that our findings do not imply a causal relationship between our variables. Instead, we must elaborate on these types of statements by sharply and precisely stating what our findings mean and what they do not mean. This is especially so when we are speaking directly to members of the media or others who may not have extensive training and background in research methodology.

When we use social media to disseminate our research findings, we must also be extremely cautious in how we report our findings, including that we must avoid using misleading, attention-grabbing titles when offering links to papers. We must keep in mind the potential misinterpretation and misapplication of our findings by first understanding research methodology and secondly, by being aware of the way in which our reports and sound bites may be interpreted.

9.4 Conclusion

I sincerely hope that I have achieved my goal of providing a broad overview of research to assist non-statistician researchers who may wish to design their own studies as well as those with a need to become good consumers of research. I did not set out to reproduce any of the many wonderful existing treatises in statistics or research design. Rather, I have endeavored to summarize in one location and in outline/primer format, the facets of research from ethics, design, analysis, interpretation, reporting, and dissemination of findings. I hope I have managed to do this sufficiently and without causing undue boredom or hypersomnolence except to those currently suffering from a lack of sleep.

   With the tools provided in the preceding chapters, you will hopefully feel more confident in reading, interpreting, and sharing research findings. Likewise, the review of research design and analysis will ideally enable you, if you are an aspiring researcher, to create a basic research design and to sit down with a statistician consultant to plan your study and later to analyze your data once they have been collected. In the best of all worlds, this primer will also serve as a handbook to which you can refer on occasion when reviewing research or when preparing your own analyses.

Appendix A

Data Sets and Databases

While data are being collected during a study, they must be recorded some-where. Depending on the type of study, data will initially be recorded in any number of different formats. These may include questionnaires completed on paper or online, observational data recorded by research assistants, or data culled from medical records and entered into a program or onto a paper form. At some point, the data will be transferred to a data set.

A data set is simply a recording of the study data (e.g. scores, demographic information for each participant). Data sets are generally recorded and stored electronically, though that has not always been the case (remember slide rulers and calculators?). Data may be entered by hand into an electronic file such as a spreadsheet or text document. Data can also be entered directly into a statistics package (e.g. SAS®, SPSS®, R, Stata® – reviewed in Appendix B). Finally, data can be recorded into a database (e.g. REDCaps, SQL, Access). I will review these database systems after I talk more about data sets.

A.1. Contents of a Data Set

A data set will include every piece of information you have about each partici-pant or subject with the exception of identifying information. Depending on your institution's policies, you may have a separate list with each person's iden-tifying information along with an alphanumeric identification code with the identification code included in the data set. That way, if you need to add infor-mation about a participant to your data set, you will have a means of identify-ing who they are so that you can go back to the original materials or records in order to obtain the correct information. In drug trials, this may be important if you identify a problem and the participant needs to be identified and con-tacted. Again, how this is managed will depend on your institution's policies.

Content in a typical data set in clinical or educational research will include an identification code, gender, age, and other demographics such as education,

Research Design and Analysis: A Primer for the Non-Statistician, First Edition. Leslie D. Rosenstein.
© 2019 John Wiley & Sons, Inc. Published 2019 by John Wiley & Sons, Inc.

socioeconomic status (SES), ethnicity, race, and hand preference (if applicable). In addition, there will be data related to any experimental conditions, such as membership in the treatment group or the control group. There will also be data for any dependent variables, such as scores on tests, lab results, success/failure, survival duration, ratings, etc.

There are different formats for recording and storing data, though I prefer to have a separate row for each participant or subject. Some longitudinal statistical analyses lend themselves better to having data with multiple lines for each participant, with each row representing a different time period. However, statistical software packages, such as SAS, can transform data from the one row format to the alternative format in a few short steps.

Below are examples of data sets. I have included the lines of information identifying the variables or data points in each row. Each line that ends with a semicolon is called a statement in SAS, and these statements included below comprise the Data Step. The actual data set is simply the matrix of numbers and alphanumeric descriptors (Table A.1).

Table A.1 Sample data set with SAS data step.

Data Totally_Madeup;
Input gender $ Handpref $ height;
Cards;
Male RH 67
Male RH 65
Male RH 68
Male RH 64
Male RH 62
Male LH 72
Male LH 73
Male LH 71
Male LH 77
Male LH 78
Female RH 67
Female RH 63
Female RH 64
Female RH 67
Female RH 64
Female LH 68
Female LH 62
Female LH 65
Female LH 67
Female LH 64
;

In this first data set, which I have aptly named, "Totally_Madeup," each row, or card, represents one participant in the study. The first variable is an alpha-numeric variable indicating the participant's gender. The second variable is also alphanumeric, and indicates the hand preference, with RH for right-handed and LH for left-handed. The third variable for each participant is the person's height in inches.

In Table A.2, I have included the Data Step as well as the data for 40 partici-pants. The first variable is the participant's ID, while the remaining three pieces

Table A.2 A sample data step.

Data motivation_madeup;			
Input ID Motivation Days_Off Stress;			
Cards;			
1	92	18	8
2	46	17	7
3	100	19	2
4	94	16	3
5	60	10	20
6	22	1	60
7	84	12	20
8	26	10	70
9	57	16	50
10	33	9	70
11	36	10	73
12	5	0	95
13	63	13	40
14	98	18	2
15	63	15	43
16	16	2	73
17	70	16	28
18	66	15	34
19	70	16	32
20	18	5	82
21	61	12	39
22	84	17	27
23	79	16	31
24	67	13	40
25	27	7	73
26	17	2	82

(Continued)

Table A.2 (Continued)

27	23	10	85
28	77	13	21
29	38	8	64
30	22	7	82
31	11	1	93
32	83	16	24
33	22	6	85
34	13	4	75
35	64	12	62
36	57	9	60
37	54	9	71
38	79	18	75
39	98	20	59
40	44	9	60
;			

of data include scores on a motivation questionnaire, days of paid time off (PTO) for a given period, and scores on a questionnaire measuring self-reported stress. In this mock study, the researchers are interested in the relationship between motivation, stress, and PTO. This data set does not include any demographic information, and there is no particular group membership such as Treatment Group versus Control.

Table A.3 includes a data set from a longitudinal or within-subjects study in which depression scores are measured before treatment, after treatment, and at follow-up. Half the participants received the intervention (treatment), and half remained on the waitlist. The first variable in each row is an alphanumeric variable indicating whether the person was in the treatment or control (wait-list) group. The second and third variables are numeric demographic variables indicating age and then years of education. The remaining three variables are the dependent variable, depression score, at each of the three time points, before the treatment period, immediately following the treatment (or waitlist) period, and at follow-up at a later date.

Depending on the statistical software package you plan to use, you may be able to enter the data into the data set as above, with only a space between variables to indicate that the next character is the next data point. With some statistical packages, you may have to carefully input each variable into specified columns.

The data sets in Tables A.1–A.3 are relatively small with a small-ish number of participants and few variables. Entering their data into a simple spreadsheet (like Excel), a document or text editor, or even directly into the statistical package is relatively easy. With large data sets, particularly data sets accumulated over time, it is probably more manageable to use a database program. Table A.4 includes

Table A.3 Data from a within-subjects study.

```
Data sample_repeated;
Input Group $ Age Ed Time1 Time2 Time3;
Cards;
Patient 25 12 44 20 21
Patient 28 14 45 17 19
Patient 21 13 65 15 17
Patient 39 11 41 11 11
Patient 38 14 44 13 10
Patient 27 9 50 11 9
Patient 50 13 55 25 22
Patient 61 12 49 14 14
Patient 47 11 35 7 9
Patient 31 11 39 5 7
Control 26 13 44 44 44
Control 28 14 45 46 49
Control 20 14 65 64 67
Control 33 12 41 40 43
Control 38 13 47 49 50
Control 41 15 52 52 52
Control 34 12 58 50 60
Control 43 17 49 50 49
Control 26 12 34 40 34
Control 24 11 39 40 39
;
```

data from a large study ($N = 1000$) with 16 independent and dependent variables. This is a data set I would not want to type or enter into a simple spreadsheet. Short of making someone else suffer through rote data entry like that, I would probably want to create a form for data entry using a database program.

A.2.   Missing Data and Data Entry Errors

One thing you may notice in Table A.4 are the dots, or periods, in place of data for participants with ID numbers 9, 10, 12, and 13. Those dots indicate that the data are missing for those variables and those participants. Missing data are not uncommon in research. Either tests are not administered for various reasons, or the participant drops out of the study when a particular measure is obtained. The dots, or periods, let the statistical software package know to read that variable for that participant as empty or missing. Another way that

Table A.4 Larger data set with 1000 participants and 16 variables.

Data sample_neuropsych;
Input ID Gender $ Age Origin $ Education Diagnosis $ RDS ROCopy ROIR RORec HVLTTot HVLTdr HVLTidr HVLThits HVLTfp TMTA TMTB;
Cards;

ID	Gender	Age	Origin	Education	Diagnosis	RDS	ROCopy	ROIR	RORec	HVLTTot	HVLTdr	HVLTidr	HVLThits	HVLTfp	TMTA	TMTB
1	M	54	Mex	9	WNL	6	35	28	22	22	9		10	1	35	51
2	M	65	US	0	MCI	7	32	20	20	16	6		12	3	48	82
3	M	39	Mex	5	ID	4	29	18	19	15	5		11	2	51	221
4	F	58	Mex	9	ADD	5	35	17	16	14	5		12	5	29	68
5	F	71	Chile	16	PD	7	34	16	16	16	5		11	1	33	79
6	F	68	US	12	MCI	5	22	14	18	17	6		10	1	45	91
7	F	74	Cuba	3	AD	5	19	16	18	10	4		8	1	40	300
8	M	75	Cuba	12	WNL	7	28	14	18	18	10		11	0	24	41
9	F	67	US	15	MCI	5	24	12	19	14	6		.	.	37	97
10	F	80	Mex	6	WNL	5	29	2	18	21	9		.	.	35	63
11	F	52	Chile	2	WNL	7	19	14	19	19	10		8	2	37	79
12	M	66	US	5.5	MCI	6	35	23	.	15	6		11	1	43	187
13	M	78	US	4	MCI	5	32	17	.	19	7		12	0	55	135
14	M	72	Mex	9	WNL	6	31	19	19	20	11		11	1	29	47
15	F	58	Cuba	5	Depr	4	18	5	11	14	4		6	11	61	69
16	M	68	Mex	2	AD	5	34	8	17	12	2		0	0	51	241
17	F	76	US	0	Depr	5	30	15	17	13	5		9	2	55	68

18	M	71	Mex	4	MCI	6	27	8	20	14	5	8	3	41	198
19	M	64	Us	4	AD	5	35	19	21	17	6	12	1	67	285
20	F	52	US	9	Depr	5	28	5	17	19	6	6	2	52	69
21	M	62	Chile	12	MCI	5	30	17	19	20	5	9	0	41	157
22	F	31	US	9	MCI	5	22	14	17	19	5	8	1	43	141
23	F	63	Mex	20	AD	3	19	2	14	14	3	8	4	78	299
24	F	68	US	0	ID	5	23	7	18	12	3	9	0	93	300
25	F	59	US	12	AD	8	26	9	17	14	2	9	6	41	257
*	*	*	*	*	*	*	*	*	*	*	*	*	*	*	*
*	*	*	*	*	*	*	*	*	*	*	*	*	*	*	*
*	*	*	*	*	*	*	*	*	*	*	*	*	*	*	*
1000	M	80	Chile	6	MCI	5	36	24	21	14	5	8	2	44	181

* Signify that there are more lines of data not included.

missing data can be denoted is by formatting your variables with column indicators. For instance, the "Input ..." line in the table above is the Input statement, and tells the statistical package, in this case SAS, the names of the variables, and their order. I could follow each variable name with the column numbers in which those data are found if in fact I format the data with the same variables always aligned in the exact same columns. Then, if a column is blank in a particular row, that datum is read as empty or missing.

So, when you enter this many pieces of data, how do you know you have not made errors? Good question! And there is an easy answer. With most statistical analysis packages, you can run a program to print the values for each variable in columns (in SAS, you could use Proc Print as one option). You will then have a nice print out of values for each variable. If any of those values does not match the characteristics of the variable, then you know you have a problem. For instance, if an alphanumeric variable shows up as having a numerical value, or if a numeric variable that should not have decimals has decimals, you know something is off in your data entry. It is usually fairly straightforward with a little sleuthing to figure out where the error occurred. Sometimes it is as simple as an omitted space or an extra space that throws off the entire data set.

As I noted earlier, it can be quite tedious to enter data into a large data set with simple typing. More importantly, it is too easy to make errors in entering data, including that spaces may be omitted, and characters may be entered incorrectly, with either type of mistake potentially throwing off the reading of the entire data set by the statistical software package. Therefore, with larger data sets, it may be preferable to use a database program for data entry.

A.3. Database Programs

There are many database programs available on the market. A few such programs are Oracle Database, SQL Server, My SQL, Microsoft Access, and REDCap. Database managers are to me, as a nonprogrammer, complex tools for completing a range of tasks from simple to grand, and they allow the user (e.g. corporations) to integrate different sources of related data. I am limiting this discussion here to a review of the simpler aspects of less-complex desktop-type database managers. Among other tasks, these programs can be used to create data sets for researchers.

Data Sets created using a desktop database manager can be set up like a table into which a researcher or their staff enters data, much like a spreadsheet. These programs also offer the option of creating a form with prompts for each variable with blanks into which the variable values can be entered (e.g. Age: 22).

If you work in an institutional setting, you may have access to staff who can assist you with setting up your database and preparing it for data entry. If you are not in such a setting, there are affordable database programs that are pretty simple to learn and use. In a Lifewire article, the top five desktop database managers reviewed in 2018 were: Microsoft Access 2016, Filemaker Pro 15, LibreOffice Base (a free open-source program based on Apache Open Office), Corel Paradox 10, and Brilliant DataBase 10 (Chapple, 2018).

In the following paragraphs, I provide a brief description of Microsoft Access 2016, which is readily available to any of us who already use Microsoft Office Suite. If you are in an institutional setting and have support available and prefer to have someone else set up your database, you will probably want to go with the database that your administrative staff and experts recommend. After the description of Access 2016, I briefly discuss REDCap, which is available at a large number of academic institutions around the world.

A.3.1   Microsoft Access

As mentioned, Microsoft Access is readily available to most of us. If you have a Microsoft Office suite, you probably have Microsoft Access. In addition to becoming more "Access'ible," it has also become more user friendly and easier to learn over the years since it was first introduced in the early 1990s. I bought Access software several years ago, but never used it, partly because the program seemed cumbersome for a beginner such as me. That is no longer the case. The basic steps to setting up a database with Access are fairly straightforward, and if you want to sharpen your skills for more flexibility and to take advantage of more of its utilities, you can find tutorials online and in the bookstore.

Microsoft Access is one of those database managers that allows you to choose whether to use old-fashioned straightforward data entry, inputting data directly into a table, or to create a form to be filled in for each observation/record (e.g. participant or subject). Either option is easy to use. You do not need to have any experience with database management or computer science. I have listed the steps to creating a very simple database using Microsoft Access below in Table A.5. This list will give you a very basic form for entering data – I recommend consulting with an Access aficionado (if you are not already one) to create anything more complicated or stylistic. You can also find all sorts of videos and tutorial online as noted earlier.

In Table A.6, you can see the data set I started creating using Microsoft Access. I created a form for easy entry as described in Table A.5.

So, once you have your data set created using a database program like Access, what do you do with it? There are lots of options, and these will depend on the statistical package you choose to use. If you are going to run simple statistics, you could export the Access file into an Excel file where you can analyze your data. There are also ways to import Access databases into statistical software packages such as SAS. If you are not comfortable learning the required steps to

Table A.5 Steps for creating a basic Microsoft Access database with forms.

1. Open the Microsoft Access software.

2. Select the labeled "Blank Desktop Database" icon.

3. Select the "Create" icon after typing in the name of your new database.

4. Make sure the "Fields" tab is selected from the top bar.

5. Start creating your variable names by selecting the tab, "Create and Add" next to "ID" at the top of the blank chart. This will bring up a drop-down menu that allows you to select the characteristic of the first variable you want to enter (e.g. Short Text for alphanumeric, Number for numeric).

6. If your first variable is alphanumeric, select, "Short Text" and then type the name of that variable (e.g. "Group," "Gender," "Medication").

7. Tab to the next "Create and Add" and select the characteristic of your next variable (Short Text, Number) and then type the name of that variable (e.g. "Age," "Baseline Score," "Height," "Hemoglobin A1C").

8. Keep repeating these steps until you have created all of your variables (i.e. Fields).

9. Save your table using the "File" tab at the top left of the screen, using "Save As" to give the table a name in the folder you desire to use for storage.

10. To create a form for easier data entry, select the name of your table under, "Tables" on the left side of the screen.

11. Select the "Create" tab at the top of the screen and then select, "Form" from the Create bar.

12. Now, save your form using the File → Save As tabs near the top left again and you are ready to enter data using Forms.

13. Select the corresponding Form from the left under "Forms" by double clicking.

14. To enter data, place your cursor over the first blank (which will show the word, "New") and tab down to the second blank. Don't make the mistake I made the first (several) tries by attempting to enter an ID number; Access won't let you. Just tab down to the second blank and enter the appropriate value (e.g. the person's age if the variable is "Age," the treatment group if the first variable is "Group"). Once you start typing the value of that second variable, the ID number will magically appear. By the way, the ID number created by Access is the record number. You can add your own unique identifier by creating a new variable, such as "Study ID."

15. Keep tabbing down and entering the appropriate values into each blank. If you keep tabbing at the end of a record (participant or subject), Access will automatically take you to the next record.

16. When you get tired, bored, or have somewhere else to be, simply save your work by selecting the save icon at the top left. Fortunately, in case your computer ever crashes, Access actually saves every time you complete entry for a new record.

17. To add records later, look for a teeny tiny icon toward the bottom left (one day Microsoft might make this more obvious) that is next to the arrows pointing to the right next to the record counter. When you drag your cursor over that tiny icon, it will read, "New (blank) record." Select the icon, and you are ready to start entering data again. If you don't see the icon, try going back to the previous record first. You do that by selecting the arrow pointing left next to the record counter at the bottom.

18. To view your data as a data set (data sheet), simply double click on the name of your file under "Tables" on the left.

Table A.6 Data set created with Microsoft Access.

		Practice Access Database Form Treatment				
ID	Group	Gender	Age	Baseline BDI	Post-treatment BDI	Follow-up BDI
1	Control	Male	45	47	25	23
2	Treatment	Female	37	63	50	47
3	Control	Male	52	60	37	27
4	Treatment	Female	34	39	22	21
5	Control	Male	50	55	50	56
6	Treatment	Female	39	62	29	26
7	Treatment	Male	52	47	19	14
8	Control	Female	41	59	49	60
9	Control	Female	29	45	39	47
10	Treatment	Female	32	41	23	25
11	Control	Female	28	49	41	53
12	Control	Male	28	43	39	44
13	Treatment	Male	27	45	22	14
14	Treatment	Male	35	41	17	15

do that, or you don't have a consultant to do that for you, you can always export the data table into a text file (see Table A.7) or an Excel file using the export icon under the "External Data" tab at the top of the Access screen. When you export to a text file, you will have the option of managing the column width, that is, the width of each variable. The width, by the way, will be denoted in units that are equivalent to the number of characters. I would recommend adding one unit, or character, to the width of each given variable so you will have a nice space between variable values for each record.

Once you have a text file, you can cut and paste the contents directly into your statistical software package, such as SAS® (which I will describe in Appendix B); see Table A.8. As noted above, though, you can also import data sets, including Access and Excel data sets, into your statistical software program – that is probably the ideal method, though the cut and paste method that we have all learned pretty well by now works, too. In Stata, you can cut and paste from an Excel file directly into the data editor.

In Table A.9, you can see the output created using SAS with the data step illustrated in Table A.8. Remember, that data step was written using the pasted text file exported from the original Access Database I created. If I can

Table A.7 Text file exported from Access database.

---

1 Control Male 45 47 25 23
2 Treatment Female 37 63 50 47
3 Control Male 52 60 37 27
4 Treatment Female 34 39 22 21
5 Control Male 50 55 50 56
6 Treatment Female 39 62 29 26
7 Treatment Male 52 47 19 14
8 Control Female 41 59 49 60
9 Control Female 29 45 39 47
10 Treatment Female 32 41 23 25
11 Control Female 28 49 41 53
12 Control Male 28 43 39 44
13 Treatment Male 27 45 22 14
14 Treatment Male 35 41 17 15

Table A.8 SAS data step using pasted data exported from Access as a text file.

```
Data Practice_Using_Access;
Input ID Group $ Gender $ Age BDI_Baseline BDI_PostTreatment BDI_Followup;
Cards;
1 Control Male 45 47 25 23
2 Treatment Female 37 63 50 47
3 Control Male 52 60 37 27
4 Treatment Female 34 39 22 21
5 Control Male 50 55 50 56
6 Treatment Female 39 62 29 26
7 Treatment Male 52 47 19 14
8 Control Female 41 59 49 60
9 Control Female 29 45 39 47
10 Treatment Female 32 41 23 25
11 Control Female 28 49 41 53
12 Control Male 28 43 39 44
13 Treatment Male 27 45 22 14
14 Treatment Male 35 41 17 15
;
```

---

produce data using a procedure like Proc Print, then I can also run analyses on the same data.

To summarize the above steps, I started by creating a form in Access and then entering data using the new form. This resulted in the creation of a data set. I then

Table A.9 SAS Proc Print output using pasted data from text file.

Obs	ID	Group	Gender	Age	BDI_ Baseline	BDI_ PostTreatment	BDI_ Followup
1	1	Control	Male	45	47	25	23
2	2	Treatment	Female	37	63	50	47
3	3	Control	Male	52	60	37	27
4	4	Treatment	Female	34	39	22	21
5	5	Control	Male	50	55	50	56
6	6	Treatment	Female	39	62	29	26
7	7	Treatment	Male	52	47	19	14
8	8	Control	Female	41	59	49	60
9	9	Control	Female	29	45	39	47
10	10	Treatment	Female	32	41	23	25
11	11	Control	Female	28	49	41	53
12	12	Control	Male	28	43	39	44
13	13	Treatment	Male	27	45	22	14
14	14	Treatment	Male	35	41	17	15

Source: Created with SAS® software. Copyright® 2018, SAS Institute Inc., Cary, NC, USA. All Rights Reserved. Reproduced with permission of SAS Institute Inc., Cary, NC.

exported the data set to a text file, which I then cut and pasted into my SAS program. I could have also exported the data set from Access to an Excel file. Alternatively, I could have used SAS or another program to import the data set directly from Access by entering code to tell SAS® (or another program) where to find the data set (i.e. the folder and file name). The beauty of data entry, data sets, databases, and statistical analyses is that there are so many options, and you can pick the one that makes the most sense to you and that is most readily available.

A.3.2   REDCap

REDCap is a powerful tool for creating surveys and data sets. The acronym, REDCap, stands for Research Electronic Data Capture. The system was developed at Vanderbilt University, and has been made available to members of a consortium of thousands of institutions around the world. It can only be used by those institutions receiving it from Vanderbilt, and it is intended to be used for research in a noncommercial manner. Generally, institutions with access to REDCap have Information Technology (IT) staff to support its use within the institution. If your institution has REDCap, the IT staff will likely help you set up your database and prepare you and your research personnel for the data entry. You can find out if your institution already has access to REDCap by searching the Project REDCap website (REDCap, 2017).

Some of the major relevant features of REDCap according to Vanderbilt's own website (Vanderbilty University, 2018) include: the ability to securely build online surveys and databases, the ability to export data to statistical software packages, the ability to create paper versions of data collection forms for collecting data off-line, and the ability to set validation rules to avoid errors in data entry. There are several websites available online with details about using REDCap, including webpages available at many of the institutions belonging to the community. If you have determined that your institution has REDCap, you can do a quick online search with your institution's name along with the title, "REDCap," and you will probably find helpful information about how to access the program and where to find support on your campus.

A.4.   Database and Data Set Summary

Just to recap, when you conduct a study, you will generally collect data. Those data will be recorded and ultimately entered into a data set for creating reports and conducting statistical analyses. When your study is small, with a small number of participants or subjects and a small number of variables, the quickest and easiest thing to do may be to type the data points into your favorite text editor or spreadsheet. You can then either import or cut and paste the text into a statistical software package. Alternatively, if you really like playing with numbers (and no one is judging here), you can get out your old calculator and hammer away at the numbers by hand. If you don't trust your big thumbs or that little calculator any more (I know I don't), or if you are not certain about the exact formulas or whether you have violated assumptions, you are probably better off using a statistical analysis software program (which I will review in Appendix B).

When you have copious amounts of data, with several observations (human participants or animal subjects), then you might start thinking about using a database to get your data entered. Databases have the added benefit of being useful for creating data sets for future archival studies, and especially exploratory studies. If you are practicing in a health field or education, you are probably creating or viewing data every day that could potentially be used to answer a research question in the future. If you have approval of your institution's research review board, and you have the resources (personnel and/or time), it would be great to enter those data into a database on an ongoing basis for future study. Down the road, you may have a research question that lends itself well to those data you have been recording all along. With multiple providers contributing data over months or years, you could ultimately have a very large sample size for some excellent analyses. Ideally, you will be able to anticipate the types of variables that you will one day be interested in studying in such an exploratory or retrospective manner, and you will have created variables and recorded the corresponding values across observations.

Whether you decide to create a simple data set, or to use a database manager to enter and store data for later analyses, it is important to plan ahead in terms of how you will analyze the data and using what statistical software package (or slide ruler for the old fashioned at heart). By planning ahead, you can be certain that your data will be complete and in a format that can be analyzed the way you intend. If you have access to a statistical consultant, someone who spends their days conducting and considering statistical analyses, I would highly recommend that you sit down with that person at the earliest point possible in your research – ideally before you have finalized your design and certainly before you have started collecting data. In addition to tweaking your design and guiding you with your data collection, that person can assist you in preparing your data set or setting up a database.

Appendix B

Statistical Analysis Packages

B.1.   SAS®, SPSS®, R, and Stata®

So, you have collected your data and entered them into a data set or database. Now, you are ready to rock and roll with your statistical analyses. To me, this is the most exciting part of a research project; those moments of anticipation as you wait while the statistical package runs your analyses to see if your results are statistically significant and in the predicted direction. All those months, if not years, of work come down to this one moment in time (cue the music).

Which statistical software package you opt to use may depend on availability. I personally like to use SAS, and it is available with a license purchased by my institution. It is also available online free for individuals associated with a qualifying academic institution vis-à-vis the free SAS University Edition. Another commonly used statistical package available at many institutions and businesses is the Statistical Package for the Social Sciences (SPSS). SAS and SPSS have much in common, and often the choice of using one or the other depends on how a researcher was raised and with access to which package. The beauty of both lies not just in their power and utility but also in the availability of resources online. With a quick online search, it is very easy to find instructions and guidance from the SAS and SPSS publishers, as well as from aficionados who have created online tutorials, videos, and sample programs.

SAS is much more interactive than when it was first developed, but it has always relied on what I consider real-world language. In other words, the statements used to tell SAS how to read your data and how and which analyses to perform are similar to natural language. Currently, SAS can be run in a Windows environment in which you can enter (i.e. type) code, which is just computerese for a list of statements, or instructions, and data. You have the option of entering data into the code, or you can point SAS to data stored outside of the program in a data set or database. As you type statements, a drop-down menu typically pops up to remind you of your options for that step.

Research Design and Analysis: A Primer for the Non-Statistician, First Edition. Leslie D. Rosenstein.
© 2019 John Wiley & Sons, Inc. Published 2019 by John Wiley & Sons, Inc.

More recently, SAS Studio has evolved as a tool in which you can use drop-down menus to specify the procedure, location and name of the data set, models (if applicable), and options. To locate the drop-down menu-style programming, you will look in the Tasks and Utilities Folder in the navigation pain of SAS Studio. The Statistics programs and the Power and Sample Size applications are located in the Tasks subfolder. Notably, SAS Studio is available with the free SAS University Edition if you are associated with a qualifying academic institution. You can find the download for the Free University Edition with a simple online search. SAS Studio is actually a web-browser programming environment (SAS Institute, 2017).

SPSS works slightly differently in that it is available in a format that includes practically all drop-down menus, so you don't have to rely as much on memory or those magically appearing pop ups of SAS. As a SAS user, it seems that SPSS has the advantage of not requiring a lot of free recall of steps, while SAS seems to have more flexibility in how you run an analysis. As noted above, the relatively new-ish SAS Studio does offer drop-down menus and increased interactive functions. SAS also seems to provide more detail and information (e.g. tables, plots, warnings, and tests of violations of assumptions) without having to specifically ask for those. For every advantage I argue that SAS has, though, I know my SPSS-using colleagues can make a counter argument. Again, I really think your preference for one over the other (SAS or SPSS: Ginger or Mary Ann) has more to do with which package you grew up using as a student and then as a researcher. After I briefly review two more statistical analysis packages, I will offer some examples of mock data analyses with SAS.

A third statistical package that has come into wide use in recent years is R. R is an open-source package and basically free for anyone to use. Many students use R in their undergraduate statistics courses, both in the sciences and in the business fields. While I refer to R as a "statistical package," the producers of R label it an "environment" in which statistics, and other operations, can be performed (R Foundation, 2018).

The origin of R is complex, but really interesting, especially if you are a history buff with interest in software development and freedom rights. To learn more about this history, I recommend visiting the R Foundation website as well as the website of the GNU (*g'noo*) Project (Free Software Foundation, 2018). Basically, the GNU Project provides programs that people are free to run, adapt, improve, and distribute. The project was founded, or announced, in 1983 by Richard Stallmann from MIT. R has stemmed from the GNU Project, and has been and continues to be developed and improved by a collaboration of users and experts.

As with SAS and SPSS, there are many easily accessible resources for learning to use and understand R. A nice introduction to R and how to start using it is, *A Handbook of Statistical Analyses Using R* by Brian Everitt and Torsten Hothorn available on the World Wide Web (Everitt & Hothorn, 2008). Another valuable, and free, website is *Statistical Tools for High-Throughput Data Analysis* (Kassambara, 2017) which has step-by-step guides for creating a data set, uploading any necessary R software, importing the data into R, and conducting statistical analyses.

In recent years, a new tool has evolved through the collaborative R programming community, and that is called R Studio. R Studio is basically an amped up, easier to use environment for R. R Studio is organized with multiple windows to view simultaneously and with drop-down menus. For a really good demonstration of using R in R Studio, I recommend the video by Mike Marin and Ladan Hamadani, *What Is R Studio and Why Should You Download It?* (2016).

Most intriguing to me is the revelation that R can be used together with both SPSS and SAS. SPSS can be used with R in an integrated fashion. This is accomplished through SPSS using the SPSS Statistics Integration Plug-in for R (IBM Software Business Analytics, 2013). Similarly, R can be called from within SAS using SAS IML/Studio. Robert Muenchen (2012) provides a good overview of how to take advantage of the many strengths of R while relying on either SAS or SPSS as your platform.

Perhaps not as commonly used by us lay people, Stata seems to be a favorite statistical software package among my more technically skilled statistician colleagues. I can see why because there are many good arguments for its use. As pointed out by Jeff Meyer of the *Analysis Factor* (Meyer, 2008–2018), Stata is relatively inexpensive, like SPSS you have the option of writing code or using simple drop-down menus with an easy-to-use help menu, it is flexible, and it can generally be used to run any statistical analysis that any other package can run. Meyer (2008–2018) points out more somewhat technical advantages of using Stata as well.

Christopher Baum at Boston College has a nice introductory text to Stata available online (Baum, 2011). From the start, Dr. Baum notes that if you use Stata to analyze a data set with a large number of variables, you will probably need to upgrade from the standard version. As with SAS and SPSS, Stata has online manuals available, and there is actually a large online community of Stata users. Thus, it should not be too difficult to find answers to any problems you encounter if you choose to use Stata. Stata also includes data-management tools that allow you to view and edit your data.

In the preceding paragraphs, I have reviewed just a few available statistical software packages, and these are the ones you will commonly hear or read about when considering behavioral and clinical research. Each has its own merits, and most of these that I have reviewed probably have more in common than not. In the following paragraphs, I provide examples of analyses using SAS. Again, SAS is the package I have the most experience with and the one that I feel most comfortable using, but others will prefer one or more of the alternatives. The point of the following exercises and demonstrations is to make the statistical analysis process less abstract for those who have never conducted their own analyses.

B.2. Sample Statistical Analyses Using SAS

In the following paragraphs, I present examples of data analyses using SAS. In this first example, I include each component of the analysis, including the cards, the log, the output data, and the results. Here, I am considering whether there are differences between a group of males and females in their height,

Table B.1 Sample SAS program for conducting an analysis of variance.

```
Data Made_up;
Input Gender $ Hand $ Height;
Cards;
Male RH 67
Male RH 65
Male RH 68
Male RH 64
Male RH 62
Male LH 72
Male LH 73
Male LH 71
Male LH 77
Male LH 78
Female RH 67
Female RH 63
Female RH 64
Female RH 67
Female RH 64
Female LH 68
Female LH 62
Female LH 65
Female LH 67
Female LH 64
;
Proc GLM;
Class Gender Hand;
Model height = Gender Hand Gender*Hand/effectsize;
Means Gender Hand Gender*Hand;
```

Source: Created with SAS® software. Copyright® 2018, SAS Institute Inc.,
Cary, NC, USA. All Rights Reserved. Reproduced with permission of SAS
Institute Inc., Cary, NC (2018).
Note: These data are fictional.

whether there are difference in the heights of the left- and right-handers, and whether there is an interaction between gender and hand preference. First, I enter the following fictional data and statements into the SAS Program Code window (see Table B.1).

To run the analysis program I just created, I select the icon of the person running (clever, right?) that is located at the top of the window in SAS Studio. I wait a few seconds and then I am presented with three windows, each viewable by selecting its respective tab. These three windows are: the Output Data, the Log, and the Results. The Output Data page shows all of my variables and their values. The Log notes any errors in my program or any warnings about the analyses – errors and warnings will be highlighted in different colors, and you will know you have problems, if any, before you even read the text. Finally, the Results page includes the results of the analyses along with any graphic presentations of the data. See Tables B.2–B.4.

In this example, I opted to run a procedure called GLM, or General Linear Models. I could have also run a procedure called ANOVA. In this sample program, I included the data steps that tell SAS about the variables – their names, whether they are numeric or alphanumeric (denoted with a "$"), and the order in which they appear in each row of data. Those statements that make up the data step are followed by the actual data set. I could have, alternatively, written

Table B.2 Sample SAS Log from an analysis of variance.

---

1 OPTIONS NONOTES NOSTIMER NOSOURCE NOSYNTAXCHECK;

72

73 Data Made_up;

74 Input Gender $ Hand $ Height;

75 Cards;

NOTE: The data set WORK.MADE_UP has 20 observations and 3 variables.

NOTE: DATA statement used (Total process time):

real time 0.00 seconds

cpu time 0.01 seconds

96;

97 Proc GLM;

98 Class Gender Hand;

99 Model Height = Gender Hand Gender*Hand/effectsize;

100 Means Gender Hand Gender*Hand;

101

102

103 OPTIONS NONOTES NOSTIMER NOSOURCE NOSYNTAXCHECK;

Source: Created with SAS® software. Copyright® 2018, SAS Institute Inc., Cary, NC, USA. All Rights Reserved. Reproduced with permission of SAS Institute Inc., Cary, NC (2018).

Table B.3 Sample SAS output data for an analysis of variance.

Obs	Gender	Hand	Height
1	Male	RH	67
2	Male	RH	65
3	Male	RH	68
4	Male	RH	64
5	Male	RH	62
6	Male	LH	72
7	Male	LH	73
8	Male	LH	71
9	Male	LH	77
10	Male	LH	78
11	Female	RH	67
12	Female	RH	63
13	Female	RH	64
14	Female	RH	67
15	Female	RH	64
16	Female	LH	68
17	Female	LH	62
18	Female	LH	65
19	Female	LH	67
20	Female	LH	64

a statement to tell SAS to get the data from a file stored on my computer. In that case, the actual data would not appear in my program code.

Following the data entry, I wrote procedure statements that told SAS which analysis to run and which options I wanted included. For instance, I asked for means to be printed for males and females, for right- and left-handers, and for the four subgroups (right-handed males, left-hand males, right-handed females, and left-handed females). I also asked for the effect sizes to be printed.

In the Output Data window, SAS showed me how it read my data – this is one point where I could check to make sure my data were read correctly, with nothing entered in the wrong column, with values that don't make sense, or with extra spaces, etc. Finally, the Results window presents the results of the

Table B.4 Sample SAS results for an analysis of variance (selected portions).

The GLM Procedure

Class-Level Information		
Class	Levels	Values
Gender	2	Female Male
Hand	2	LH RH

Number of Observations Read	20
Number of Observations Used	20

Dependent Variable: Height

Source	DF	Sum of Squares	Mean Square	F Value	Pr > F
Model	3	308.4000000	102.8000000	16.72	<0.0001
Error	16	98.4000000	6.1500000		
Corrected total	19	406.8000000			

Source	DF	Type III SS	Mean Square	F Value	Pr > F
Gender	1	105.8000000	105.8000000	17.20	0.0008
Hand	1	105.8000000	105.8000000	17.20	0.0008
Gender*Hand	1	96.8000000	96.8000000	15.74	0.0011

Source		Noncentrality Parameter			
		Min Var Unbiased Estimate	Low MSE Estimate	95% Confidence Limits	
Gender		14.1	12.0	2.83	42.6
Hand		14.1	12.0	2.83	42.6
Gender*Hand		12.8	10.9	2.37	39.9

(Continued)

Table B.4 (Continued)

Source	Partial Variation Accounted For			
	Partial eta-square	Partial omega-square	95% Confidence limits	
Gender	0.5181	0.4476	0.1239	0.6806
Hand	0.5181	0.4476	0.1239	0.6806
Gender*Hand	0.4959	0.4243	0.1058	0.6659

Level of gender	Level of hand	N	Height	
			Mean	Std Dev
Female	LH	5	65.2000000	2.38746728
Female	RH	5	65.0000000	1.87082869
Male	LH	5	74.2000000	3.11448230
Male	RH	5	65.2000000	2.38746728

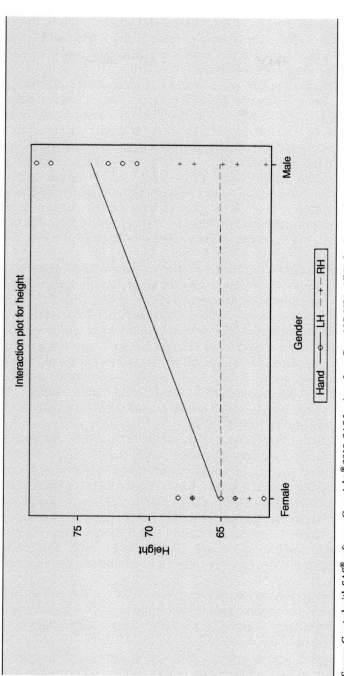

analyses I requested. In Table B.4, I have only included parts of the results. I omitted some of the tables of means, some of the F tests, and some of the graphics and effect sizes, but this gives you a general idea of what a SAS program looks like along with the results.

In these results, you can see that there is a significant effect of gender, such that males and females differ in their height with males being taller than females ($F_{[1,16]} = 17.20, p = 0.0008$). There is also a significant effect of hand preference with left-handers being taller than right-handers, overall ($F_{[1,16]} = 17.20, p = 0.0008$). Finally, there is a significant interaction between gender and hand preference ($F_{[1,16]} = 15.74, p = 0.0011$). Specifically, from the table of means and the graph, you can see that it is the left-handers that account for the difference in height between the two genders. Alternatively, you could say that it is the males that account for the difference in height between the right- and left-handers. As you can see in the table of means and the graph, the left-handed males are taller than all three of the other groups.

In this next example, I analyzed data from a fictional study in which I administered an IQ test to 33 swimmers selected at random from a group of swimmers attending a national swim meet. As can be seen in Table B.5, I named my data set "IQ." I named my two variables, "ID" and "IQ." I then entered the ID's and IQ scores for each of the 33 swimmers followed by the procedure statements. Here, I used a procedure called, "Proc TTEST." Since this was a single sample t test, I entered the value of the null hypothesis in the analysis options ($H_0 = 100$, denoted in SAS as h0 = 100) – my null hypothesis is that the mean IQ score of the group of swimmers is not different than the known mean IQ score of the population ($\mu = 100$). I also requested the 95% confidence intervals by including the option, "alpha = .05".

As was noted in the discussion of this study in Chapter 4, the results of the analysis reveal a statistically significant result ($t = 10.11, p < 0.0001$). This allows us to reject the null hypothesis of no difference between our group of swimmers and the general population in IQ scores. In other words, the swimmers were found to have significantly different IQ scores than the general population. This is further corroborated by the confidence interval, presented as the 95% confidence limits of 113.6 and 120.4. This means that the true mean for swimmers, accounting for sampling error, probably falls somewhere between 113.6 and 120.4, and this range does not include the null hypothesis mean of 100.

SAS results also typically include graphs and plots, which will show the distribution of scores as well as a graphic representation of the confidence interval in this case. With practice and skill, it is possible to set and manage the parameters of the graphs to suit your needs.

In the following fictional analysis illustrated in Table B.6, I have conducted a Chi-square test using frequency data from a data set that contains additional types of data that can be analyzed separately. In this fictional study, I was

Table B.5 Sample *t* test with confidence intervals run using SAS.

Code
Data IQ;
Input ID IQ;
Cards;
1 91
2 97
3 102
4 107
5 112
6 112
7 112
8 115
9 115
10 115
11 116
12 116
13 116
14 116
15 117
16 117
17 117
18 117
19 117
20 118
21 118
22 118
23 118
24 119
25 119
26 119
27 122
28 122
29 122
30 127
31 132
32 137
33 143
;
Proc ttest h0 = 100 alpha = 0.05;
Var IQ;
Run;

(Continued)

Table B.5 (Continued)

Results

The TTEST Procedure

Variable: IQ

N	Mean	Std Dev	Std Err	Minimum	Maximum
33	117.0	9.6566	1.6810	91.0000	143.0

Mean	95% CL Mean		Std Dev	95% CL Std Dev	
117.0	113.6	120.4	9.6566	7.7657	12.7727

DF	t Value	Pr > |t|
32	10.11	<0.0001

Source: Created with SAS® software. Copyright ® 2018, SAS Institute Inc., Cary, NC, USA. All Rights Reserved. Reproduced with permission of SAS Institute Inc., Cary, NC (2018).

interested in knowing whether a rural and a city hospital differ in the likelihood that a patient complaining of depressive symptoms will be diagnosed with Major Depression versus Generalized Anxiety Disorder (GAD) or Depressive disorder, Not Otherwise Specified.

As can be seen in the *Statistics for Table of Diagnosis by Hospital* section, the Chi-square is significant (Chi-square = 12.76, p = 0.0017), so the null hypothesis that there is no association between hospital and diagnosis can be rejected. That is, from these results I can conclude that the two hospitals differ in the likelihood with which patients are diagnosed with Major Depression versus GAD versus Depressive Disorder, Not Otherwise Specified. With the data collected and included in the above data set, I can also conduct separate analyses to consider whether the likelihood of one diagnosis or another is impacted by factors such as age, education, and depression scores.

In this next example outlined in Table B.7, I have conducted a regression analysis with a relatively small sample of individuals seen in a primary care clinic. I was interested in learning whether the degree of depression could be predicted on the basis of age, education, and hours exercised each week.

As can be seen in Table B.7, it appears that my model was statistically significant (F = 27.63, $p < 0.0001$) with exercise contributing significantly and negatively to depression scores, meaning more exercise per week was associated with lower depression scores. Age and education did not appear to impact

Table B.6 Sample Chi-square analysis run using SAS.

Code
Data Chi_Square_Example;
Input ID Age Education Hospital $ Diagnosis $ Depression_Score;
Cards;
01 29 12 City Major_Depression 29
02 35 13 City Major_Depression 35
03 28 11 Rural Depressive_DO_NOS 41
04 31 13 Rural GAD 37
05 39 12 Rural Major_Depression 31
06 41 13 City Depressive_DO_NOS 28
07 45 14 City Major_Depression 31
08 37 10 Rural GAD 35
09 32 12 City Depressive_DO_NOS 25
10 35 12 Rural GAD 33
11 39 13 Rural Major_Depression 45
12 25 16 City Major_Depression 28
13 44 13 City Depressive_DO_NOS 19
14 32 12 Rural GAD 25
15 37 15 City Major_Depression 25
16 29 12 City GAD 19
17 31 14 Rural Depressive_DO_NOS 29
18 40 12 Rural Depressive_DO_NOS 17
19 41 14 City Major_Depression 40
20 38 12 Rural Depressive_DO_NOS 40
21 27 11 Rural GAD 37
22 29 13 City Major_Depression 35
23 31 12 City Depressive_DO_NOS 31
24 35 12 Rural GAD 35
25 49 13 Rural Depressive_DO_NOS 30
26 47 13 City Major_Depression 30
27 35 12 Rural GAD 32
28 31 13 City GAD 21
29 28 14 City Depressive_DO_NOS 23
30 35 12 Rural GAD 28
31 27 13 Rural Depressive_DO_NOS 32
32 32 12 Rural GAD 35
33 41 14 City Major_Depression 29
34 29 12 City Depressive_DO_NOS 25
35 41 14 City Major_Depression 32

(Continued)

Table B.6 (Continued)

36 29 13 City Major_Depression 31
37 28 12 Rural Depressive_Disorder_NOS 35
38 35 15 Rural GAD 31
39 28 14 Rural Depressive_Disorder_NOS 37
40 35 13 City Major_Depression 31
;
Proc Freq;
Tables Diagnosis*Hospital/chisq plots = freqplot(type = dot scale = percent);
Run;

Results
The FREQ Procedure

Table of Diagnosis by Hospital			
Diagnosis	**Hospital**		
Frequency **Percent** **Row Pct** **Col Pct**	**City**	**Rural**	**Total**
Depressi	6	8	14
	15.00	20.00	35.00
	42.86	57.14	
	30.00	40.00	
GAD	2	10	12
	5.00	25.00	30.00
	16.67	83.33	
	10.00	50.00	
Major_De	12	2	14
	30.00	5.00	35.00
	85.71	14.29	
	60.00	10.00	
Total	20	20	40
	50.00	50.00	100.00

Table B.6 (Continued)

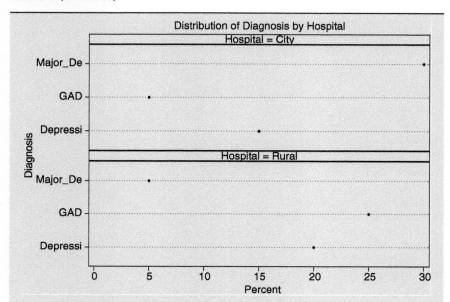

Statistics for Table of Diagnosis by Hospital

Statistic	DF	Value	Prob
Chi-square	2	12.7619	0.0017
Likelihood ratio Chi-square	2	14.0336	0.0009
Mantel–Haenszel Chi-square	1	5.0143	0.0251
Phi coefficient		0.5648	
Contingency coefficient		0.4918	
Cramer's V		0.5648	

Sample size = 40

Table B.7 Sample multiple regression analysis run using SAS.

Code
Data Fictional_Depression_Study;
Input ID Depression_Score Age Education Exercise_Hours_Week;
Cards;
1 0 25 12 6
2 4 30 13 5
3 4 35 12 5
4 8 20 11 5
5 8 25 12 4
6 8 30 13 4
7 8 35 14 5
8 13 20 14 4
9 13 25 13 5
10 13 30 12 3
11 13 35 13 4
12 13 40 11 4
13 13 35 12 4
14 13 25 12 5
15 17 20 12 4
16 17 40 11 3
17 17 25 13 3
18 17 30 14 4
19 17 35 13 4
20 17 40 12 4
21 17 30 11 3
22 17 25 12 4
23 17 30 10 4
24 17 35 14 3
25 21 25 12 4
26 21 20 13 3
27 21 40 12 3
28 21 35 11 3
29 21 30 13 2
30 21 25 14 3
31 26 25 11 3
32 26 40 12 2
33 26 35 13 2

Table B.7 (Continued)

```
34 30 25 12 2
;

Proc Reg Data = Fictional_Depression_Study Corr Simple;
Model Depression_Score = Age Education Exercise_Hours_Week / Spec;
output out = fitdata rstudent = jackknife;
Proc Univariate Data = fitdata normal;
var jackknife;
run;
Proc Reg Data = Fictional_Depression_Study;
model Depression_Score = Age Education Exercise_Hours_Week;
output out = fitdata cookd = cooksd h = hat;
run;
quit;
Proc Print Data = fitdata;
where hat ge 8/37;
run;
Proc Print Data = fitdata;
where cooksd ge 4/37;
run;
```

Results and Tests of Assumptions

The REG Procedure

Number of Observations Read	34
Number of Observations Used	34

			Descriptive Statistics		
Variable	Sum	Mean	Uncorrected SS	Variance	Standard Deviation
Intercept	34.000 00	1.000 00	34.000 00	0	0
Age	1 020.000 00	30.000 00	31 950	40.909 09	6.396 02
Education	419.000 00	12.323 53	5 199.000 00	1.073 98	1.036 33
Exercise_ Hours_Week	126.000 00	3.705 88	500.000 00	1.001 78	1.000 89
Depression_ Score	535.000 00	15.735 29	9 935.000 00	45.958 11	6.779 24

(Continued)

Table B.7 (Continued)

Correlation				
Variable	Age	Education	Exercise_Hours_ Week	Depression_ Score
Age	1.0000	−0.1143	−0.2130	0.0804
Education	−0.1143	1.0000	0.0069	−0.0737
Exercise_ Hours_Week	−0.2130	0.0069	1.0000	−0.8470
Depression_ Score	0.0804	−0.0737	−0.8470	1.0000

The REG Procedure
Model: MODEL1
Dependent Variable: Depression_Score

Number of Observations Read	34
Number of Observations Used	34

Analysis of Variance					
Source	DF	Sum of squares	Mean square	F value	Pr > F
Model	3	1 113.606 71	371.202 24	27.63	<0.0001
Error	30	403.010 94	13.433 70		
Corrected total	33	1 516.617 65			

Root MSE	3.665 20	R-Square	0.7343
Dependent mean	15.735 29	Adj R-Sq	0.7077
Coeff Var	23.292 87		

Parameter Estimates					
Variable	DF	Parameter Estimate	Standard Error	t Value	Pr > |t|
Intercept	1	47.740 93	9.131 70	5.23	<0.0001
Age	1	−0.121 20	0.102 78	−1.18	0.2476
Education	1	−0.528 49	0.619 83	−0.85	0.4006
Exercise_Hours_ Week	1	−5.897 87	0.652 54	−9.04	<0.0001

Table B.7 (Continued)

Test of First and Second Moment Specification		
DF	**Chi-square**	**Pr > ChiSq**
9	12.87	0.1688

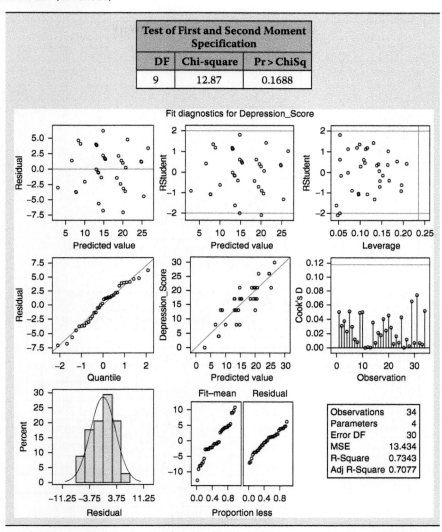

Fit diagnostics for Depression_Score

Observations	34
Parameters	4
Error DF	30
MSE	13.434
R-Square	0.7343
Adj R-Square	0.7077

(Continued)

Table B.7 (Continued)

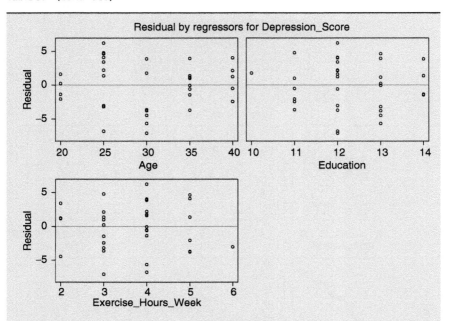

The UNIVARIATE Procedure
Variable: jackknife (Studentized Residual without Current Obs)

Moments			
N	34	Sum Weights	34
Mean	−0.001 655 4	Sum Observations	−0.056 282 7
Std deviation	1.020 158 3	Variance	1.040 722 96
Skewness	−0.287 488 6	Kurtosis	−0.709 288 5
Uncorrected SS	34.343 950 9	Corrected SS	34.343 857 7
Coeff variation	−61 627.073	Std error mean	0.174 955 71

Basic Statistical Measures			
Location		Variability	
Mean	−0.001 66	Std Deviation	1.020 16
Median	0.178 63	Variance	1.040 72
Mode		Range	3.894 48
		Interquartile Range	1.509 47

Table B.7 (Continued)

Tests for Location: Mu0 = 0				
Test	**Statistic**		**p Value**	
Student's t	t	−0.00946	Pr > |t|	0.9925
Sign	M	1	Pr > = |M|	0.8642
Signed rank	S	11.5	Pr > = |S|	0.8475

Tests for Normality				
Test	**Statistic**		**p Value**	
Shapiro–Wilk	W	0.969855	Pr < W	0.4575
Kolmogorov–Smirnov	D	0.110874	Pr > D	>0.1500
Cramer–von Mises	W-Sq	0.053339	Pr > W-Sq	>0.2500
Anderson–Darling	A-Sq	0.335789	Pr > A-Sq	>0.2500

Quantiles (Definition 5)	
Level	**Quantile**
100% Max	1.809836
99%	1.809836
95%	1.427577
90%	1.188244
75% Q3	0.616731
50% Median	0.178627
25% Q1	−0.892738
10%	−1.316847
5%	−1.992405
1%	−2.084647
0% Min	−2.084647

Extreme Observations			
Lowest		**Highest**	
Value	**Obs**	**Value**	**Obs**
−2.08465	10	1.18040	20
−1.99240	5	1.18824	14
−1.61780	6	1.35444	9
−1.31685	29	1.42758	31
−1.07794	2	1.80984	25

depression scores. The total model accounted for approximately 71% of the variance in depression scores (Adjusted R-Square = 7.077).

Fortunately, particularly given the relatively small sample size, the analysis appears to meet the major assumptions of regression. Inclusive, the data and model appear to have homoscedasticity of errors (Chi-square = 12.87, p = 0.169) as well as normality of the errors (e.g. W = 0.97, p = 0.458). I have also checked for outliers, and all appears well there, too. Finally, the three predictor variables do not correlate excessively with one another. For a good review of tests of assumptions in regression using SAS, I recommend *Regression Diagnostics* (Hill, 2009) which is actually a set of lecture notes from a course taught by Dr. Hill, and still available online as of the writing of this primer. There are additional similar resources to be found online with a simple search using the terms *assumptions*, *regression*, and the name of your preferred statistical software package.

B.3. Summary

In this section, I have presented four different types of statistical analyses conducted using SAS. Fortunately, there are many examples available online that can also be reviewed. Within the SAS publications online, it is quite easy to find examples of any statistical procedure available. Similarly, there are copious examples demonstrating the use of SPSS, Stata, R, and other tools for performing various analyses. Most of these can be found using a simple search using your favorite search engine. Others are found in various publications as noted earlier in Section B.1.

Appendix C

Helpful Statistics Resources

In this Appendix, I am including a listing of helpful resources for the novice statistician as well as for the researcher who just needs to check with experts on questions about analyses. In Section C.1, I have listed some of my favorite textbooks as well as a couple goodies recommended by my trusted colleagues. In Section C.2, I have included websites that may be useful to have on hand. These lists are by no means exhaustive, and others in your circle of research colleagues may have their favorites.

C.1.   Statistics Textbooks

1. *Introductory Statistics for the Behavioral Sciences, 4th Edition* (Young & Veldman, 1981): This is my all-time favorite introductory statistics text, probably because Robert Young, the first author, was my original stats mentor. I like the way the authors intermingle humor with statistics, as well as the way they use understandable, real-world-type examples in their explanations and exercises. I have used this book to teach undergraduates, and found that the nonthreatening and non-pretentious style was beneficial for my students who arrived to the first day of class with pre-existing math anxiety. This book is quite old now (did I mention I studied under Dr. Young when I was just in Kindergarten?), so it may be hard to find, but there are copies out there.

2. *Advanced Statistics in Research: Reading, Understanding, and Writing Up Data Analysis Results* (Hatcher, 2013): This is another textbook that integrates humor and real-world types of research questions to explain statistics concepts, in this case at a more advanced level than the Young and Veldman text. This book is also much newer, and can be found in your favorite online bookstore. Dr. Hatcher additionally has helpful online exercises and materials that coordinate with the book. One thing I appreciate about this book is

Research Design and Analysis: A Primer for the Non-Statistician, First Edition. Leslie D. Rosenstein.
© 2019 John Wiley & Sons, Inc. Published 2019 by John Wiley & Sons, Inc.

the use of a similar set of studies addressed with different questions and analyses.

3. *Statistics, Third Edition* (Hays, 1981): This is a comprehensive statistics text book intended for the serious learner who really wants to understand how statistics work at a mathematical level. My copy is well-worn, and I still pull it out periodically as a reference. There are newer editions available (though they are getting old as well), but as of the writing of this book, the Third Edition is still available and for next to nothing (as in less than $2 – that is not a typo: two bucks). You can get the Fifth Edition if you are okay with the expense, but the third edition has plenty of material that does not change much over time.

4. *Multivariable Analysis: A Practical Guide for Clinicians* (Katz, 1999, 2006, 2011): This is a statistics textbook recommended by one of my trusted colleagues, a professor and prolific researcher. The title alone says it all – a *practical* guide. This book reportedly reviews multivariate statistics in a manner that can be comprehended by clinicians without a strong background in multivariate analyses. In a nutshell, multivariate analyses are complex analyses in which more than one outcome or dependent variable is considered at a time.

5. *Statistical Rules of Thumb* (van Belle, 2008): This is another one recommended by my colleague referred to above. This book even includes discussions of observational studies and evidence-based medicine. You can see from online reviews that individuals in the research industry, including statisticians, find this book useful as a reference tool as well as a review of a wide range of statistical topics.

6. *Basic and Clinical Biostatistics* (Dawson & Trapp, 2004): A third statistics book recommended by the same trusted colleague. This one is described as an introduction to biostatistics, and includes review and discussion of published studies as well as accompanying data sets and statistical software (on CD-ROM) to use for practice. There is also a chapter on survey research, as well as further review of multivariate analyses.

7. *Statistical Reasoning in Psychology and Education* (Minium & King, 2003): This one was recommended by another of my favorite colleagues. It has gone through multiple iterations since she first purchased it; she notes that the authors made everything so clear. Throughout the book there are "points of controversy" (always fun), including some discussion of meta-analysis.

8. *Applied Statistics for the Behavioral Sciences* (Hinkle, Wiersma, & Jurs, 2002): Another favorite colleague recommended this textbook, which is now in its Fifth Edition. This is an introductory overview of inferential statistics that can be used for learning and/or reference. There are discussions of SPSS® included, with sample printouts and explanations for interpretation of the results.

C.2.   Statistics Websites

Below is a listing of some of the websites I frequently consult when searching for an answer to a question or when trying to find a good explanation of a concept to share with my students. In fact, I often share pages from these sites with my students. In addition to being very useful, these sites are free to use and easy to access as long as you have power and a device connected to the World Wide Web. Some of these sites do include subscription options and allow you to send questions to the authors or to join conversations with others interested in the topics. Whether you are a Sheldon Cooper of statistics or a Forrest Gump, these websites can provide useful material to improve your knowledge about a statistics topic. These are not listed in any particular order.

1. www.theanalysisfactor.com: *The Analysis Factor* (The Analysis Factor, 2008–2018b)
2. www.statisticshowto.com: *Statistics How To: Statistics for the rest of us* (Glen, 2018)
3. https://statistics.laerd.com: *Laerd Statistics* (Lund & Lund, 2018)
4. www.statisticssolutions.com: *Complete Dissertation* (Statistics Solutions, 2018)
5. http://www.statsoft.com/Textbook: *Electronic Statistics Textbook* (StatSoft, 2013)
6. https://rcompanion.org/rcompanion/e_02.html: *An R Companion for the Handbook of Biological Statistics* (Mangiafico, 2015)

In addition to the above websites devoted to statistics topics, there are many more websites to be found with simple online searches. I will often Google a particular statistics topic or question with Google returning a wealth of resources. I tend to open websites with addresses ending with the ".edu" extension as these are often materials posted by statistics professors for their students. In addition to static websites, these simple search results often include videos which can often be invaluable.

Glossary

Abstract A brief summary typically included at the beginning of a research report for submission to a journal for publication, containing a statement of the purpose of the study, a brief description of the participants/subjects and methods, a summary of the results, and a brief discussion of the meaning and importance of the findings moving forward.

ANCOVA Analysis of covariance, which is a statistical procedure to test for differences between two or more groups while controlling for the effects of one or more extraneous variables.

ANOVA Analysis of variance, which is a statistical procedure to test for differences between two or more groups.

ANOVA-RM Repeated measures analysis of variance, which is a statistical procedure to test for differences across time or repeated measurement within a group; it may also be used to test for differences between groups across time.

Assumptions (statistical) Basically, a set of requirements for using a statistical technique, such as the type of data, the shape of the distribution, and the equality of variances across groups.

Attrition Also referred to as experimental mortality, meaning participants or subjects are lost from a study for various reasons.

Beecher report A 1966 review of several studies funded by the government that were felt to have caused harm or potential harm to participants (Beecher, 1966).

Between-subjects design A research design in which groups are compared.

Between–within subjects design A research design in which groups are compared over time or across repeated measurements.

Bimodal distribution A frequency distribution of scores in which there are two separate concentrations of scores resulting in the presence of two modes.

Blinding A strategy in experimental research in which the investigator and/ or the participant are not informed of the participant's experimental

Research Design and Analysis: A Primer for the Non-Statistician, First Edition. Leslie D. Rosenstein.
© 2019 John Wiley & Sons, Inc. Published 2019 by John Wiley & Sons, Inc.

condition (e.g. real drug versus placebo) in order to prevent placebo effects or bias in assessment.

Blocking A strategy used in research design to assess the effects of more than one independent variable at a time. Groups of interest are divided on the basis of the main independent variable, and then they are further subdivided on the basis of a second (or more) independent variable. The second (third, fourth, etc.) variable(s) is used to manage extraneous variance, but there may also be an interest in learning about possible interactions between the main variable and the blocking variable(s).

Bonferroni *t* test Also referred to as the Dunn test, a test used for planned or post hoc comparisons between pairs of means, with the α-level used for significance divided by the number of comparisons made – the Bonferroni correction – to counteract the increased risk of Type I errors that occurs with multiple comparisons.

Carryover effects When participants in a study experience more than one research condition (i.e. in a within-subjects design), their reactions to the first condition may still be present when they are exposed to the second condition; also referred to as order effects.

Chi-square A statistical test used to compare the relative frequencies of one outcome versus other outcomes across groups, to test for an association between one grouping and another, or to test whether observed frequencies are consistent with expectation based on theory or known population frequencies (i.e. goodness of fit).

Cochrane risk bias tool A tool composed of a set of criteria against which to judge studies and their designs when considered for inclusion in a meta-analysis.

Complex comparison A type of planned comparison in which a single group's mean or a combined groups' mean can be compared to another group's or combined groups' mean (e.g. the combined mean of Groups 1 and 2 can be compared to the combined mean of Groups 3, 4, and 5).

Confidence interval A measure of precision, providing a range in which the true descriptive statistic, or parameter, of the population (e.g. mean score) may fall based on our sample findings.

Confound In regression analysis, a variable that correlates with both the predictor and the criterion variable such that a spurious relationship between the predictor and criterion is created.

Continuous variable A variable measured on a ratio or interval scale that has an infinite number of values, such as 1.135, 1.136, 1.17, 1.9, 2, and everything in between.

Correlation A measure of the strength and direction of the relationship between two variables measured on interval or ratio scales.

Covariate An extraneous variable that can be quantified, typically as a value on an interval or ratio scale, and that correlates significantly with the dependent variable in a statistical analysis; used in Analysis of Covariance (ANCOVA) to statistically control for the effects of the extraneous variable.

Criterion Another term for dependent variable, particularly in studies in which causality can not be addressed.

Crossover design A within-subjects research design in which each participant or subject crosses over from one condition to the other(s); each participant is exposed to each level of the within-subjects, independent variable.

Data set Data from a research study (e.g. scores, experimental group membership, demographic information for each participant, etc.), often recorded and stored electronically.

Database A computer software tool that can be used for creating, updating, editing, and managing data sets.

Declaration of Helsinki A set of principles for ethical research originally adopted in 1964 by the World Medical Association and updated several times since.

Dependent *t* test Also referred to as a matched pairs *t* test, a statistical test used to compare two sets of related scores measured on an interval or ratio scale; these related scores can include a repeated measure with the same group of participants or subjects tested at two points.

Dependent variable The outcome measure in a statistical analysis, such as the result of a test or the characteristic being measured and compared across groups.

Discriminant function analysis A statistical procedure used to predict membership in one level or another of a categorical variable based on a set of predictor variables, all of which are measured on an interval or ratio scale.

Discussion section In a research paper, the last section before the references, and generally composed of four elements – a brief summary of the findings, a presentation of any limitations of the study, a summary of the importance of the findings and how they may apply going forward, and suggestions for future research.

Distribution In inferential statistics, the layout of scores by frequency (i.e. a frequency plot), illustrating the central tendency and the spread of scores.

Double Latin Square A research design that facilitates assignment of participants to levels of the main independent variable (e.g. treatment modality) in a crossover, within-subjects design while controlling for order effects.

Dunn test Also referred to as the Bonferroni *t* test, a test used for planned or post hoc comparisons between pairs of means, with the α-level used for significance divided by the number of comparisons made – the Bonferroni correction – to counteract the increased risk of Type I errors that occurs when multiple comparisons are made.

Dunnett test A planned comparison conducted when an analysis of variance reveals significant group differences; in this case, each of the experimental groups is compared to the control group and not to one another.

Effect size The strength, or magnitude, of a statistical finding (e.g. the difference, the relationship).

Experimental design A research design in which the independent variable (e.g. the treatment arm, the condition, etc.) is manipulated, or created, participants are randomly assigned to levels of the independent variable, and extraneous influences are controlled and equal at all levels of the independent variable.

Experimental mortality Also referred to as attrition, meaning participants or subjects are lost from a study for various reasons.

Experimenter bias The researcher (unintentionally) influences the outcome of the study by their own biases, which may impact administration of the intervention or assessment of the response.

External validity The degree to which a study's results generalize outside of the study setting and to other groups or populations.

Family-wise error rate The risk, or probability, of Type I errors that occurs with multiple comparisons in a statistical analysis.

Forest plot In a meta-analysis, a table or figure that illustrates each study's effect size and confidence interval relative to the summary effect size and its confidence interval.

Freedom to withdraw The concept that a research participant may discontinue participation in a study at any time without consequence (e.g. without their care or other benefits being impacted).

Frequency data Counts of the number of members in a particular nominal group or category.

Funnel plots In a meta-analysis, used to assess whether there appears to be a publication bias by plotting the effect size for each of a set of studies against their sample sizes. If the plot does not have the appearance of an inverted funnel, smaller studies with small effect sizes have likely not been published at the same rate as small studies with large effect sizes.

Gamma Goodman and Kruskal's measure of association between two sets of rankings – useful when there are many tie scores/ranks in the data (i.e. multiple individuals with the same scores/ranks on one or both of the variables).

Graeco-Latin square A Latin square design used to assign research participants or subjects to groups (e.g. intervention groups) when there are three nuisance variables to be managed in a study designed to analyze differences between the groups.

History In research, events that occur over the course of time during which an experiment is being conducted, and which can impact the validity of the research findings if not managed (such as with the use of a control group).

Homogeneity of variance One of the assumptions of several parametric statistical procedures – the variance, or spread, of the scores in each of the groups is approximately equal.

Hyper Graeco-Latin square A Latin square design used to assign research participants or subjects to groups (e.g. intervention groups) when there are *four* nuisance variables to be managed in a study designed to analyze differences between the groups.

I^2 A test of the degree of heterogeneity of effect sizes across studies included in a meta-analysis.

Independent samples *t* test A statistical procedure to compare two different independent groups on a dependent variable that is measured on a ratio or interval scale.

Independent variable The conditions (e.g. intervention, category) the researcher is interested in evaluating in a study.

Informed consent The concept of research participants agreeing to participate in a study only after being made aware of all aspects of the study as well as their right to decline or, later, withdraw from the study.

Institutional Review Board An agency in an institution that is responsible for reviewing, approving, and monitoring research, including with respect to ensuring the safety and wellbeing of human research participants and animal subjects.

Interaction In a statistical analysis comparing groups with more than one independent variable, the interaction effect indicates the degree to which there is a differential effect of one independent variable across levels of the other independent variable(s).

Internal validity The degree to which the design of a study and the control of extraneous variables allow for the deduction of a causal link between the independent and dependent variables.

Interval data Numbers or values on a scale with an equal distance between each number and the next; similar to ratio data except there is no absolute zero value, so ratios cannot be calculated to reflect fractions.

Introduction In a research paper, the first section following the abstract, with three major components – an explanation of the importance of the topic and why it should be studied, a presentation of the relevant literature, and the study hypotheses with a brief description of the methods used to test them.

Kendall's tau A measure of the association between two sets of rankings that is useful with smaller samples or when there are potential errors in the data coding – it is based on concordance and discordance between pairs of scores/ranks.

Kruskal–Wallis test A nonparametric statistical test sometimes referred to as a one-way ANOVA for ranks, and used to compare two or more independent samples when the data violate the assumption of normality and the sample size is not large, or when the data are ordinal in nature.

Latin square design A special type of multifactorial between–within subjects design that allows the researcher to control for two extraneous variables (called nuisance variables) by treating them as between-subjects independent variables while managing the number of participants required to complete the study.

Least Significant Difference test A post hoc test following the finding of a significant analysis of variance, used to compare each group to one another using the least difference between means that is significant based on a *t* test; tends to have an inflated Type I error rate.

Leptokurtic Positively kurtotic; a distribution of scores that appears tall and thin due to the presence of a few outliers in the extremes and more scores found in the middle.

Main effect In analysis of variance, the effect of an independent variable standing alone.

MANCOVA Multivariate analysis of covariance, a statistical procedure used to test for differences between two or more groups while controlling for the effects of one or more extraneous variable(s), and with two or more dependent variables considered together.

Mann–Whitney *U* test Wilcoxon rank sum test, a nonparametric statistical procedure similar to a *t* test for independent samples, but used when the assumptions of the *t* test are not met and the samples are small; based on rankings, so it can also be used with ordinal data.

MANOVA Multivariate analysis of variance, a statistical procedure used to test for differences between two or more groups, with two or more dependent variables considered together.

Matched pairs *t* test Also referred to as a dependent *t* test, a statistical test used to compare two sets of related scores measured on an interval or ratio scale; these related scores can include a repeated measure with the same group of participants or subjects tested at two points.

Matching A strategy to manage extraneous variables in a research study that compares groups; accomplished by assigning equal numbers of individuals with the same value on the extraneous variable (e.g. ethnicity) to each level of the independent variable of interest (e.g. intervention method). Splitting twin pairs, one sibling into each group, is a special form of matching.

Materials section In the Methods section of a research paper, this subsection provides a listing and description of any measures or instruments used in the study, including information about the reliability and validity of newer tools. If a new questionnaire was developed for the study, a description of how that instrument was developed and tested is also included.

Maturation The natural changes that occur over time, including among participants or subjects in a study; a potential threat to the internal validity of a study that can be managed by having a comparison control group.

Mean The average; calculated by totaling all the scores, and dividing that sum by the number of scores.

Measurement error Anything that reduces the reliability or precision of the assessment or quantification of the dependent (outcome) variable.

Median The middle score among a range of scores that are rank ordered from the lowest to the highest score; when there is an even number of scores, or values, the median is the average of the two middle scores.

Median test Mood's median test; a special case of Chi-square test that analyzes whether the medians of two or more samples are equal.

Mediator In regression, a mediator explains the relationship between two other variables by indicating whether the relationship between the predictor and criterion can be explained as being due to a third variable (the mediator); the predictor variable impacts the value of the mediator, which, in turn, impacts the value of the criterion.

Meta-analysis A compilation of results across studies asking a specific question, and typically producing a summary effect size that ideally represents the true effect in the population being investigated.

Methods section In a research paper, this section is generally divided into a subsection describing the participants or subjects, a subsection describing the materials and procedures, and, often, a subsection describing the planned statistical analyses.

Milgram's study The Milgram (1963) experiment on obedience to authority figures; a well-known study involving deception in which participants were instructed to administer shocks to another individual under the guise of a study assessing the impact of punishment on learning.

Mixed Factorial design A between–within subjects design, also referred to as a split-plot design; includes both between-subjects variables and repeated measures.

Mixed Linear Model analysis A statistical procedure that can be used for analyzing repeated measures data; capable of handling missing data and treating time as a continuous variable.

Mode The most common value among a group of discrete values in a frequency distribution.

Moderator In regression, a variable that interacts with a predictor variable in influencing the criterion variable; moderators can enhance, buffer, or reverse the effect of the predictor variable.

Multifactorial design A research design in which more than one factor, or independent variable, is evaluated in one analysis.

Multiple regression A statistical procedure used to determine to what degree a set of predictor variables predicts a criterion variable.

Negatively skewed A frequency distribution in which the tail is longer on the left side; the mean is pulled down, toward the tail on the left, such that it is smaller than the mode, with the median somewhere between the mean and the mode.

Nesting A strategy for managing nuisance variables when it is not feasible to have a fully balanced design.

Newman–Keuls test A post hoc analysis to compare pairs of means, but performed sequentially starting with the two groups with the largest difference between their means and continuing to the next largest mean difference, and so on until the difference between a pair of means is no longer significant; this has the effect of reducing the number of comparisons.

Nominal data Names or categories, often measured in a sample in terms of frequency (i.e. frequency or number in each category).

Nonparametric statistics Statistical analysis procedures used particularly (1) with frequency and ordinal data, (2) when interval or ratio data are not normally distributed, (3) when there are outliers, or (4) when there are major differences in group variances (i.e. heterogeneity of variances).

Normal distribution A frequency distribution that is symmetric with the mode, median, and mean being equal with a fairly significant drop off in frequencies as we move away from the mode, with very few scores occurring at the extremes.

Null hypothesis The hypothesis that there is no effect in our study – such as no correlation(s) between our variables or no difference(s) between the group means.

Nuremberg code A set of ethical research principles that stemmed, in large part, from the Nuremberg trials where scientists were convicted of war crimes for the atrocities perpetrated by the Nazis, including atrocities involving human experimentation.

Odds ratio The odds of successful outcome in one group (e.g. a treatment group) relative to the odds of successful outcome in another group (e.g. a control group).

Order effects When participants in a study experience more than one research condition (i.e. in a within-subjects design), their reactions to the first condition may still be present when they are exposed to the second condition; also referred to as carryover effects.

Ordinal data Data expressed in a manner that the values can be ranked such that their order has meaning (e.g. very good, good, neutral, bad, and very bad); ranks cannot be characterized as equidistant from one another in terms of the quantity of the underlying variable being ranked.

Outliers A score or value in a frequency distribution that is unusually, or extremely, high or low compared to the rest of the values in the distribution.

Parameter A descriptive value, such as a mean or variance of a measurement, in the population.

Parametric statistics Statistical procedures that can be used to analyze data when certain assumptions are met, and these assumptions generally rely on interval or ratio data that are normally distributed.

Participants section In a research paper, the first subsection of the Methods section; includes a detailed description of the sample(s), sample size(s), major demographics of the sample(s), inclusion and exclusion criteria for participants, participant recruiting methods, assignment to groups, and power analyses conducted to determine the ideal sample sizes.

Percentiles Values that are often reported as ranks below which a particular percentage of the sample or population falls on a measure.

Phi coefficient A nonparametric statistic that provides a measure of the strength and direction of the relationship, or association, between two dichotomous variables.

Planned comparisons Planned comparisons between *prespecified* pairs of means following the finding of significant differences across groups when conducting an analysis of variance.

Platykurtic A frequency distribution that is negatively kurtotic, appearing squattier than a normal distribution due to having more scores in the tails, and not as much concentration in the middle; the tails are not composed of extreme values, or outliers, relative to the remainder of the scores.

Point–biserial correlation A measure of the direction and strength of the relationship (correlation) between one continuous variable and one binary, or dichotomous variable.

Positively skewed A frequency distribution in which the tail is longer on the right; the mean is pulled upward toward the tail, such that the value of the mean is higher than the mode, with the median midway between the two.

Poster presentation A pictorial presentation of a research study, with text and graphics mounted or printed on a medium the size of a typical poster, and usually displayed in an exhibit hall during a poster session at a conference.

Post hoc comparisons Exploratory comparisons between pairs of means following the finding of significant differences across groups when conducting an analysis of variance.

Power The probability that you will reject a null hypothesis and conclude significance when there really is an "effect" in the population (i.e. correct rejection of the null hypothesis).

Power analyses Used to estimate the power of a study; can be used to determine the sample size needed to detect a difference given a desired power, a specific effect size, and a set alpha level.

Predictor An alternate term for *independent variable*, often used when the research design does not permit conclusions about causality.

Procedures section In the Methods section of a research paper, this subsection includes a brief description of the design as well as details about how the study was conducted.

Publication bias Also referred to as the *file drawer effect*, this is the tendency for studies with nonsignificant results to be filed away and never published, particularly those with smaller sample sizes.

Q statistic Referred to as Cochran's Q Statistic, this is a statistical tool for testing for the presence of heterogeneity in effect sizes across the studies included in a meta-analysis.

Quasi-experimental A research design that meets most of the criteria of a true experiment with the exception that the independent variable, or condition of interest, cannot be manipulated because it is innate or fixed (such as hand preference, country of origin, etc.).

R An open-source online tool described as an "environment" in which statistics, and other operations, can be performed; used by many as a statistical software package.

Random assignment One of the strategies used to maximize the internal validity of a study; participants or subjects are put into one of the intervention or control groups, for example, using an unbiased (random) approach to avoid systematic, pre-existing differences between the groups.

Randomized block design In a research study, participants are first divided into groups on the basis of a quasi-experimental, extraneous variable, and then randomly assigned from their quasi-experimental groups to one of the experimental groups (i.e. conditions of interest).

Ratio data Values, expressed as numbers, measured on a scale in which there is an absolute zero value, and an equal distance between each number and the next such that a data point can be expressed as a ratio.

Raw scores Scores as reported in their original metric (e.g. pounds, inches, or years).

Reactive effects When research participants react differently than they would naturally because they are in a laboratory setting or because they are aware they are in a study; this impacts external validity.

REDCap Research Electronic Data Capture, a powerful tool developed at Vanderbilt University for creating surveys and data sets.

Reliability The precision and repeatability of a measurement tool; this includes the degree to which a tool obtains the same results across repeated measurements (e.g. test–retest reliability) and with different agents (e.g. inter-rater reliability).

Results section In a research paper, this section includes a description and the result of each of the statistical analyses conducted to test the hypotheses of the study, with both a verbal characterization of each analysis as well as the actual numerical result.

Right to withdraw In ethical research involving humans, the concept that a participant may discontinue participation at any time and without consequence to them.

SAS® A set of software tools that includes statistical software packages for conducting analyses.

Scheffe's test A post hoc or planned analysis following a significant analysis of variance, allowing for multiple comparisons of pairs of groups as well as comparisons between the means of combined groups.

Sign test A nonparametric version of *t* tests, useful when there are outliers in the data or other major violations of the assumptions of *t* tests.

Single-factor design A research design with one factor, or independent variable.

Spearman's rank-order correlation A measure of the direction and strength of the relationship between the rankings of two variables; useful when at least one of the variables is ordinal in nature.

Sphericity An assumption in repeated measures analysis of variance (ANOVA-RM), basically requiring that the variances of the difference scores between pairs of any and all combinations of the repeated measure are approximately the same.

Split-plot design A between–within subjects design, also referred to as a mixed factorial design; includes both between-subjects variables and repeated measures.

SPSS® Statistical Package for the Social Sciences, a statistical software package currently owned by IBM®.

Standard deviation A measure of the spread of scores in a frequency distribution; the square root of the variance.

Stanine A standard score with a mean of 5 and a standard deviation of 2.

Stata® A statistical software package that also includes data management tools.

Statistical significance A judgment about the likelihood of obtaining a particular statistical value, determined by assessing what the probability of obtaining that numerical value for a given sample size would be if there was no true effect; if the likelihood is determined to be below the preset alpha level (e.g. probability = 0.05), then the statistical value, or result, is described as being statistically significant.

***T* score** A standard score with a mean of 50 and a standard deviation of 10.

Test validity The extent to which the quantification of a variable, or score, actually measures the intended phenomenon; various types of validity include construct validity (convergent and divergent), content validity, criterion-related validity (predictive and concurrent), and face validity.

Tukey Honestly Significant Difference (HSD) A post hoc test following the finding of a significant analysis of variance, used to compare each group to one another; maintains the Type I error rate by adjusting alpha according to the number of comparisons, but the calculation of the differences is mathematically slightly different than the Bonferroni *t* tests.

Tuskegee experiment A study frequently cited as an unethical study that caused harm to the participants; conducted by the United States Public Health Service from 1932–1972 to monitor the natural course of syphilis even after penicillin became the accepted treatment of choice in the 1940s.

Type I error In decision making in inferential statistics, this is the error made when the null hypothesis is rejected when, in fact, it is true; the risk of a Type I error in interpreting a given single statistical result will be equal to the preset alpha level (i.e. the *p* value).

Type II error In decision making in inferential statistics, this is the error made when the null hypothesis is not rejected when, in fact, it is not true (i.e. failure to reject a false null hypothesis); the risk of a Type II error is inversely related to power, with less power increasing the risk of a Type II error.

Undue influence A research concept in which an individual may be persuaded to participate or remain in a study against their own judgment or, at the very least, with a degree of discomfort and lack of desire and true consent; the persuasion may be intentional, or it may be inadvertent due, for example, to the participant's desire to please the researcher.

Variance A measure of spread in a frequency distribution of scores; based on the average squared deviation of scores around the mean.

Wilcoxon signed-rank test Similar to the sign test, a nonparametric version of a *t* test, but it additionally analyzes the *magnitude* of the difference by considering ranks as well as signs.

Within-subjects design A research design in which all participants are assigned to one group, but the dependent variable is measured at multiple points (also known as a repeated measures design).

z-score A standard score representing the number of standard deviations that a score falls from the mean for the population when the population distribution is normal.

References

Abdi, H., & Williams, L. J. (2010). Newman-Keuls test and Tukey test. In N. Salkind (Ed.), *Encyclopedia of research design* (pp. 1–11). Thousand Oaks, CA: Sage.

Akhtar, A. (2015). The flaws and human harms of animal experimentation. *Cambridge Quarterly of Healthcare Ethics, 24,* 407–419.

alphascore. (2015). LSAT score conversion. LSAT Prep Course Online. Retrieved from https://www.alphascore.com/resources/lsat-score-conversion

American Medical Association. (1966a, November 30). Ethical guidelines for clinical investigation. Illinois Institute of Technology: Ethics Codes Collection. Retrieved from http://ethics.iit.edu/ecodes/node/6240

American Medical Association. (1966b, November 30). Ethics codes collection. Illinois Insitute of Technology. Retrieved from http://ethics.iit.edu/codes/AMA%201966.pdf

American Medical Association. (2018). Code of medical ethics: Research and innovation. Retrieved from https://www.ama-assn.org/delivering-care/code-medical-ethics-research-innovation

American Psychological Association (2010). *Publication manual of the American Psychological Association* (6th ed.). Washington, DC: Author.

American Psychological Association. (2016). Ethical principles of psychologists and code of conduct. Retrieved from http://www.apa.org/ethics/code

Anastasi, A., & Urbina, S. (1997). *Psychological testing* (7th ed.). Upper Saddle River, NJ: Prentice Hall.

Baker, L., Foster-Schubert, K., Green, P. S., Wilkinson, C. W., McTierman, A., Plymate, S., ... Craft, S. (2010). Effects of aerobic exercise on mild cognitive impairment: A controlled trial. *Archives of Neurology, 67,* 71–79.

Baum, C. F. (2011, August). Introduction to Stata. Faculty Micro Resource Center. Retrieved from http://fmwww.bc.edu/GStat/docs/StataIntro.pdf

Beck, A., Epstein, N., Brown, G., & Steer, R. A. (1988). An inventory for measuring clinical anxiety: Psychometric properties. *Journal of Consulting and Clinical Psychology, 56,* 893–897.

Research Design and Analysis: A Primer for the Non-Statistician, First Edition. Leslie D. Rosenstein.
© 2019 John Wiley & Sons, Inc. Published 2019 by John Wiley & Sons, Inc.

Beck, A. T., Steer, R. A., & Brown, G. K. (1996). *Manual for the Beck depression inventory-II*. San Antonio, TX: Psychological Corporation.

Beecher, H. (1966). Ethics and clinical research. *New England Journal of Medicine, 274*, 1354–1360.

Belluz, J. (2018, April 2). 20 Years ago, research fraud catalyzed the anti-vaccination movement. Let's not repeat history: How Andrew Wakefield's shoddy science fueled autism-vaccine fears. Retrieved from https://www.vox.com/2018/2/27/17057990/andrew-wakefield-vaccines-autism-study

Benito-León, J., & Bermejo-Pareja, F. (2013). Cognitive decline in short and long sleepers. *Journal of Psychiatric Research, 47*, 1998–2003.

Bherer, L., Erickson, K. I., & Liu-Ambrose, T. (2013). A review of the effects of physical activity and exercise on cognitive and brain functions in older adults. *Journal of Aging Research*, 1–8.

Bordens, K. S., & Abbott, B. B. (2011). *Research design and methods: A process approach* (8th ed.). New York City: McGraw-Hill.

Borenstein, M., Hedges, L., & Rothstein, H. (2007). Meta-analysis: Fixed effect vs. random effects. Comprehensive meta-analysis. Retrieved from www.Meta-Analysis.com

Brower, V. (2003). A second chance for hormone replacement therapy? *European Molecular Biology Organization Reports, 4*, 1112–1115 Retrieved from Science and Society https://www.ncbi.nlm.nih.gov/pmc/articles/PMC1326430/pdf/4-7400043.pdf.

Centers for Disease Control and Prevention. (2017a). U.S. Public Health Service Syphillis Study at Tuskegee. Retrieved from https://www.cdc.gov/tuskegee/after.htm

Centers for Disease Control and Prevention. (2017b, March 3). Measles (Rubeola). Retrieved from https://www.cdc.gov/measles/about/complications.html

Centers for Disease Control and Prevention. (2017c, August 30). U.S. Public Health Service Syphilis Study at Tuskegee: The Tuskagee Timeline. Retrieved from https://www.cdc.gov/tuskegee/timeline.htm

Centers for Disease Control and Prevention. (2018, June 27). Mumps. Retrieved from https://www.cdc.gov/mumps/outbreaks.html

Chapman, S., Aslan, S., Spence, J. S., DeFina, L. F., Keebler, M. W., Didehbani, N., & Lu, H. (2013). Shorter term aerobic exercise improves brain, cognition, and cardiovascular fitness in aging. *Frontiers in Aging Neuroscience, 5*, 1–9.

Chapple, M. (2018, July 1). Top desktop databases. Retrieved from https://www.lifewire.com/top-desktop-databases-1019622

Cheng-Tek Tai, M. (2012). Deception and informed consent in social, behavioral, and educational research (SBER). *Tzu Chi Medical Journal, 24*, 218–222.

Clark, J. C. (2006). A critique of the Women's Health Initiative Studies (2002–2006). *Nuclear Receptor Signaling, 4*, 1–10.

Cook, S. W., Edgerton, H. A., Ferguson, L. W., Krugman, M., Sargent, H. D., Super, D. E., … Hobbs, N. (1951). Ethical standards for psychology: Section 4--ethical standards in research. *The American Psychologist, 6*, 427–452.

Craft, L., & Perna, F. M. (2004). The benefits of exercise for the clinically depressed. *Primary Care COmpanion Journal of Clinical Psychiatry, 6*, 104–111.

Cruelty to Animals [39 & 40 VICT. Ch. 77.]. (1876, August 15). Retrieved from http://www.legislation.gov.uk/ukpga/1876/77/pdfs/ukpga_18760077_en.pdf

Dansfield, B., & Brightwell, B. (2013). Kruskall-Wallis ANOVA: Use and misuse. Retrieved from http://influentialpoints.com/Training/Kruskal-Wallis_ANOVA_use_and_misuse.htm

Darley, J. M., & Latané, B. (1968). Bystander intervention in emergencies: Diffusion of responsibility. *Journal of Personality and Social Psychology, 8*, 377–383.

David, F. J., Robichaud, J. P., Leurgans, S. E., Poon, C., Goldman, J. G., Comelia, C. L., ... Corcos, D. M. (2015). Exercise improves cognition in Parkinson's disease: The PRET-PD randomized clinical trial. *Movement Disorders, 30*, 1657–1663.

Dawson, B., & Trapp, R. G. (2004). *Basic and Clinical Biostatistics* (4th ed.). New York: Lange Medical Books-McGraw-Hill.

de Lench, B. (2017, August 9). CTE: Is the media scaring young athletes to death? Retrieved from https://m.huffpost.com/us/entry/us_5988d42de4b08a4c247f2503/amp

Ellis, P. D. (2010, May 30). Effect size FAQs: Research that matters, results that make sense. Retrieved from https://effectsizefaq.com/2010/05/30/can-you-show-me-how-to-do-meta-analysis-in-just-2-minutes

Everitt, B. S., & Hothorn, T. (2008). A handbook of statistical analyses using R. Retrieved from cran.r-project.org: https://cran.r-project.org/web/packages/HSAUR/vignettes/Ch_introduction_to_R.pdf

Faul, F., Erdfelder, E., Lang, A.-G., & Buchner, A. (2007). G*Power 3: A flexible statistical power analysis program for the social, behavioral, and biomedical sciences. *Behavior Research Methods, 39*, 175–191.

Field, A. P., & Gillett, R. (2010). How to do a meta-analysis. *British Journal of Mathematical and Statistical Psychology, 63*, 665–694.

Figdor, C. (2017, February 2). (When) Is science reporting ethical? The case for recognizing shared epistemic responsibility in science journalism. Frontiers in Communication: Science and Environmental Communication. Retrieved from https://www.frontiersin.org/articles/10.3389/fcomm.2017.00003/full

Finch, W. H. (2007). Performance of the Roy-Bargmann stepdown procedure as a follow up to a significant MANOVA. *Multiple Linear Regression Viewpoints, 33*, 12–22.

Fischer, B. (2006). A summary of important documents in the field of research ethics. *Schizophrenia Bulletin, 32*, 69–80.

Flaherty, D. K. (2011). The vaccine-autism connection: A public health crisis caused by unethical medical practices and fraudulent science. *Annals of Pharmacotherapy, 45*, 1302–1304.

Franco, N. H. (2013). Animal experiments in biomedical research: A historical perspective. *Animals, 3*, 238–273.

Free Software Foundation. (2018, July 14). What is GNU? GNU Operating System. Retrieved from https://www.gnu.org/home.en.html

Germany {Territory under Allied occupations, 1945-1955; U.S. Zone} (1949). Permissible medical experiments. In *Trials of war criminals before Nuremberg, Nuremberg Military Tribunals* (pp. 181–184). Washington, DC: US Government Printing Office.

Glen, S. (2018). StatisticsHowTo. Retrieved from www.statisticshowto.com

Government Publishing Office. (2018, July 19). Part 46: Protection of human subjects. Electronic Code of Federal Regulations. Retrieved from https://www.ecfr.gov/cgi-bin/retrieveECFR?gp=&SID=83cd09e1c0f5c6937cd9d7513160fc3f&pitd=20180719&n=pt45.1.46&r=PART&ty=HTML

Government Publishing Office [US]. (2016, January 1). Code of Federal Regulations, Title 9: Animals and animal products, Subchapter A: Animal welfare. Retrieved from https://www.gpo.gov/fdsys/pkg/CFR-2016-title9-vol1/xml/CFR-2016-title9-vol1-chapI-subchapA.xml

Guiney, H., Lucas, S. J., Cotter, J. D., & Machado, L. (2015). Evidence cerebral blood-flow regulation mediates exercise-cognition links in healthy young adults. *Neuropsychology, 29*, 1–9.

Haidich, A. (2010). Meta-analysis in medical research. *Hippokratia, 14*(Suppl 1), 29–37.

Hamilton, M. (1960). A rating scale for depression. *Journal of Neurology, Neurosurgery, and Psychiatry, 23*, 56–62.

Hatcher, L. (2013). *Advanced statistics in research: Reading, understanding, and writing up data analysis results.* Saginaw, MI: Shadow Finch Media.

Hayley, A., Williams, L. J., Venugopal, K., Kennedy, G., Berk, M., & Pasco, J. (2015). The relationship between insomnia, sleep apnoea (sic) and depression: Findings from the American National Health and Nutrition Examination Survey. *Australian and New Zealand Journal of Psychiatry, 49*, 156–170.

Hays, W. L. (1981). *Statistics* (3rd ed.). New York: Holt, Rinehart and Winston.

Heaton, R. K., Miller, S. W., Taylor, M. J., & Grant, I. (2004). *Revised comprehensive norms for an expanded Halstead-Reitan battery: Demographically adjusted neuropsychological norms for African American and Caucasion Adults.* Lutz, FL: Psychological Assessment Resources.

Hein, I. M., De Vries, M. C., Troost, P. W., Meynen, G., Van Goudoever, J. B., & Lindauer, R. J. (2015). Informed consent instead of assent is appropriate in children from the age of twelve: Policy implications of new findings on children's competence to consent to clinical research. *BMC Medical Ethics, 16*, 76.

Higgins, J. P., & Altman, D. G. (2008). Assessing risk of bias in included studies. In J. P. Higgins, & S. Green (Eds.), *Cochrane handbook for systematic reviews of interventions: Cochrane book series* (pp. 187–241). West Sussex: Wiley.

Higgins, J. P., Altman, D. G., Gøtzsche, P. C., Jüni, P., Moher, D., Oxman, A. D., … Sterne, J. A. (2011). The cochrane collaboration's tool for assessing risk of bias in randomized trials. *BMJ (British Medical Journal), 343*, d5928.

Hill, E. G. (2009). Regression diagnostics. Retrieved from http://people.musc. edu/~hille/2009BMTRY755_Website/LectureNotes/Diagnostics/ diagnosticsStudentNotes.pdf

Hinkle, D. E., Wiersma, W., & Jurs, S. G. (2002). *Applied statistics for the behavioral sciences* (5th ed.). Boston, MA: Houghton Mifflin.

Hoffman, J. I. (Ed.) (2015). Meta-analysis. In *Biostatistics for medical and biomedical practicioners* (pp. 645–653). London: Elsevier.

IBM Software Business Analytics. (2013, October). The power of IBM SPSS statistics and R together. Retrieved from https://www.ibm.com/ developerworks/community/files/form/anonymous/api/library/ b5bb8a42-04d2-4503-93bb-dc45d7a145c2/document/0f1a7b3b-31eb-4b62- 9c62-1daa4d163a13/media/WhitePaper-UsingRandStatisticsTogether.pdf

Jones, D. S., Grady, C., & Lederer, S. E. (2016). "Ethics in Clinical Research": The 50th Anniversary of Beecher's Bombshell. *The New England Journal of Medicine, 374,* 2393–2398.

Kalron, A., & Zeilig, G. (2015). Efficacy of exercise intervention programs on cognition in people suffering from multiple sclerosis, stroke and Parkinson's disease: A systematic review and meta-analysis of current evidence. *NeuroRehabilitation, 37,* 273–289.

Kassambara, A. ((2017). Statistical tools for high-throughput data analysis. Retrieved from http://www.sthda.com/english

Katz, M. (1999). *Multivariate analysis: A practical guide for clinicians.* New York: Cambridge University Press.

Katz, M. (2006). *Multivariate analysis: A practical guide for clinicians.* New York: Cambridge University Press.

Katz, M. (2011). *Multivariate analysis: A practical guide for clinicians.* New York: Cambridge University Press.

Kikoy, H. (2018, July 26). Japan's Dr. Mengele: Medical experiments on POW's at Unit 731. Retrieved from https://www.warhistoryonline.com/history/unit-731- prison-camp-japanese.html

Klaiber, E. L., Vogel, W., & Rako, S. (2005). A critique of the Women's Health Initiative hormone therapy study. *Fertility and Sterility, 84,* 1589–1601.

Kripke, D. F., Garfinkel, L., Wingard, D. L., Klauber, M. R., & Marler, M. R. (2002). Mortality associated with sleep duration and insomnia. *Archives of General Psychiatry, 59,* 131–136.

Kuhn, A. W., Yengo-Kahn, A. M., Kerr, Z. Y., & Zuckerman, S. L. (2016). Sports concussion research, chronic traumatic encephalopathy and the media: Repairing the disconnect. *British Journal of Sports Medicine, 51,* 1732–1733.

LaCroix, A. Z., Chlebowski, R. T., Manson, J. E., Aragaki, A. K., Johnson, K. C., Martin, L., ... Wactawski-Wende, J. (2011). Health risks and benefits after stopping the Women's Health Initiative Trial of conjugated equine estrogens in postmenopausal women with prior hysterectomy. *Journal of the American Medical Association, 305,* 1305–1314.

Lautenschlager, N., Cox, K. L., Flicker, L., Foster, J. K., Bockxmeer, F. M., Xiao, J., ... Almeida, O. P. (2008). Effect of physical activity on cognitive function in older adults at risk for Alzheimer (sic) disease: A randomized title. *JAMA*, *300*, 1027–1037.

LeBlanc, E. S., Hillier, T. A., Pedula, K. L., Rizzo, J. H., Cawthon, P. M., Fink, H. A., ... Browner, W. S. (2011). Hip fracture and increased short-term but not long-term mortality in healthy older women. *Archives of Internal Medicine*, *171*, 1831–1837.

Lehman, E., Hein, M. J., & Gersic, C. M. (2016). Suicide mortality among retired National Football League players who played 5 or more seasons. *American Journal of Sports Medicine*, *44*, 2486–2491.

Liu, Y., Wheaton, A. G., Chapman, D. P., & Croft, J. B. (2013). Sleep duration and chronic disease among US adults age 45 years and older: Evidence from the 2010 Behavioral Risk Factor Surveillance System. *Sleep*, *36*, 1421–1427.

Lobo, R. (2013). Where are we 10 years after the Women's Health Initiative. *The Jounral of Clinical Endocrinology and Metabolism*, *98*, 1771–1780.

Lund, A., & Lund, M. (2018). Laerd statistics. Retrieved from Statistics.Laerd.com: https://statistics.laerd.com/

Majumder, M. S., Cohn, E. L., Mekura, S. R., Huston, J. E., & Brownstein, J. S. (2015). Substandard vaccination compliance and the 2015 measles outbreak. *Jama Pediatrics*, *169*, 494–495.

Mangiafico, S. (2015). *An R companion for the handbook of biological statistics*. New Brunswick, NJ: Rutgers Cooperative Extension Retrieved from http://rcompanion.org/documents/RCompanionBioStatistics.pdf.

Marin, M., & Hamadani, L. (2016, August 30). What is R Studio and why should you download it? Marin Stats Lectures. Retrieved from https://www.youtube.com/watch?v=riONFzJdXcs

McCurry, J. (2015, August 13). Japan revisits its darkest moments where American POWs became human experiments. The Guardian: US Edition. Retrieved from https://www.theguardian.com/world/2015/aug/13/japan-revisits-its-darkest-moments-where-american-pows-became-human-experiments

McDonald, J. H. (2014). *Handbook of biological statistics*. Baltimore, MD: Sparky House Publishing Retrieved from http://www.biostathandbook.com/transformation.html.

McHugh, M. L. (2013). THe Chi-square test of independence. *Biochemia Medica*, *23*, 143–149.

Meyer, J. (2008–2018). Why use Stata. The Analysis Factor. Retrieved from https://www.theanalysisfactor.com/why-use-stata

Milgram, S. (1963). Behavioral study of obedience. *Journal of Abnormal and Social Psychology*, *67*, 371–378.

Minitab Express Support. (2017). Data considerations for Kruskal-Wallis Test. Retrieved from https://support.minitab.com/en-us/minitab-express/1/

help-and-how-to/modeling-statistics/anova/how-to/kruskal-wallis-test/
before-you-start/data-considerations

Minium, E. W., & King, B. M. (2003). *Statistical reasoning in psychology and education* (4th ed.). Hoboken, NJ: Wiley.

Muenchen, R. A. (2012). Calling R from other software. Retrieved from https://r4stats.com/articles/calling-r

National Academy of Sciences. (1991). Science, medicine, and animals. Retrieved from https://www.ncbi.nlm.nih.gov/books/NBK223354

National Center for Health Statistics. (2000, November 21). 2 to 20 years: Girls Stature-for-age and weight-for-age percentiles. CDC Safer Healthier People. Retrieved from https://www.cdc.gov/growthcharts/data/set2clinical/cj41c072.pdf

National Commission for the Protection of Human Subjects of Biomedical and Behavioral Research (1979). *The Belmont report: Ethical principles and guidelines for the protection of human subjects of research.* Bethesda, MD: Author.

National Insitutes of Health. (1994, March 18). NIH guidelines on the inclusion of women and minorities as subjects in clinical research. NIH Guide. Retrieved from https://grants.nih.gov/grants/guide/notice-files/not94-100.html

National Institutes of Health. (2015). Public Health Service Policy on Humane Care and Use of Laboratory Animals. Office of Laboratory Animal Welfare. Retrieved from https://olaw.nih.gov/sites/default/files/PHSPolicyLabAnimals.pdf

National Institutes of Health. (2017, December 6). NIH policy and guidelines on the inclusion of women and minorities as subjects in clinical research. Retrieved from https://grants.nih.gov/grants/funding/women_min/guidelines.htm

National Institutes of Health Office of Laboratory Animal Welfare. (2015). Public Health Sevice Policy on Humane Care and Use of Laboratory Animals. Retrieved from https://olaw.nih.gov/sites/default/files/PHSPolicyLabAnimals.pdf

National Research Council Committee on Recognition and Alleviation of Pain in Laboratory Animals (2009). *Recognition and alleviation of pain in laboratory animals.* Washington, DC: The National Academies Press.

Nirphirake, A. (2018, January 3). How can we tackle heterogeneity in meta-analysis. Retrieved from https://www.cytel.com/blog/tackling-heterogeneity-in-meta-analysis

Nunnally, J. C. (1994). *Psychometric theory* (3rd ed.). New York City: McGraw-Hill.

O'Flynn, E. (2015, April 9). US bomber crew shot down over Japan were dissected while ALIVE in horrific WW2 esperiments: Japanese university acknowledges full details of atrocity 70 years on. Retrieved from https://www.dailymail.co.uk/news/article-3028694/U-S-POWs-shot-Japan-70-years-ago-dissected-ALIVE-macabre-experiments-controversial-new-exhibition-shows.html

Pascual-Leone, A., Singh, T., & Scoboria, A. (2010). Using deception ethically: Practical research guidelines for researchers and reviewers. *Canadian Psychology, 51*, 241–248.

Penn State University. (2018). 3.3 - Multiple Comparisons. Retrieved from https://onlinecourses.science.psu.edu/stat503/node/15

Polanin, J. R., Hennessy, E. A., & Tanner-Smith, E. E. (2017). A review of meta-analysis packages in R. *Journal of Educational and Behavioral Statistics, 42*, 206–242.

Quiñones, J. (2008–Present). What would you do? *ABC News.*

R Foundation. (2018). What is R? The R Project for statistical computing. Retrieved from https://www.r-project.org/about.html

Ratey, J., & Loehr, J. E. (2011). The positive impact of physical activity on cognition during adulthood: A review of underlying mechanisms, evidence, and recommendations. *Reviews in the Neurosciences, 22*, 171–185.

REDCap. (2017). Partners. Retrieved from https://projectredcap.org/partners

Reitan, R. (1958). Validity of the trail making test as an indication of organic brain damage. *Perceptual and Motor Skills, 8*, 271–275.

Rice, T. W. (2008). The historical, ethical, and legal background of human-subjects research. *Respiratory Care, 53*, 1325–1329.

Robinson, W. M., & Unruh, B. T. (2008). The hepatitis experiments at the Willowbrook State School. In E. J. Emanuel, C. Grady, R. A. Crouch, R. K. Lie, F. G. Miller, & D. Wendler (Eds.), *The Oxford textbook of clinical research ethics* (pp. 80–85). New York: Oxford University Press.

Rossouw, J.E., Anderson, G.L., Prentice, R.L., LaCroix, A.Z., Kooperberg, C., Stefanick, M.L., ... Writing Group for the Women's Health Initiative Investigators. (2002). Risks and benefits of estrogen plus progestin in healthy postmenopausal women: Principal results from the Women's Health Initiative randomized controlled trial. *Journal of the American Medical Association, 288*: 321–333.

Sabanayagam, C., & Shankar, A. (2010). Sleep duration and cardiovascular disease: Results from the National Health Interview Survey. *Sleep, 33*, 1037–1042.

SAS Institute. (2017). SAS Studio: Fact sheet. Retrieved from https://www.sas.com/content/dam/SAS/en_us/doc/factsheet/sas-studio-107188.pdf

SAS Institute. (2018). SAS® software. Cary, NC: Author.

Schwartz, S., & Gottlieb, A. (1981). Participants' postexperimental reactions and the ethics of bystander research. *Journal of Experimental Social Psychology, 17*, 396–407.

Sheu, C.-F., & Suzuki, S. (2001). Meta-analysis using linear mixed models. *Behavior Research Methods, Instruments, and Computers, 33*, 102–107.

Society of Professional Journalists. (2014, September 6). Society of Professional Journalists Code of Ethics. Retrieved from https://www.spj.org/ethicscode.asp

Sommers, R., & Miller, F. G. (2013). Forgoing debriefing in deceptive research: Is it ever ethical? *Ethics and Behavior, 23*, 98–116.

Statistics Solutions. (2018). Statistics solutions: Advancement through clarity. Retrieved from www.statisticssolutions.com

Stats Direct. (2005–2018a). Heterogeneity in Meta-analysis. Retrieved from https://www.statsdirect.com/help/meta_analysis/heterogeneity.htm

Stats Direct. (2015–2018b). Studentized Range (Q) Distribution. Retrieved from https://www.statsdirect.com/help/distributions/studentized_range.htm

StatSoft (2013). *Electronic statistics textbook.* Tulsa, OK: Author. Retrieved from StatSoft.com: http://www.statsoft.com/textbook/

Steyn, P. (2015, October 8). Which test: Logistic regression or discriminant function analysis. Retrieved from http://www.introspective-mode.org/logistic-regression-or-discriminant-function-analysis

Sun, J., Freeman, B. D., & Natanson, C. (2018). Meta-analysis of clinical trials. In J. Gallin, F. P. Ognibine, & L. L. Johnson (Eds.), *Principles and practice of clinical research* (4th ed.) (pp. 317–327). London: Elsevier.

Taylor, L. E., Swerdfeger, A. L., & Eslick, G. D. (2014). Vaccines are not associated with autism: An evidence-based meta-analysis of case-control and cohort studies. *Vaccine, 32,* 3623–3629.

Texas A&M University. (2012). Human subjects in resarch: Investigators resources. Texas A&M University Research Compliance and Biosafety. Retrieved from https://rcb.tamu.edu/humansubjects/forms/templates

The Analysis Factor. (2008–2018a). Six differences between repeated measures ANOVA and linear mixed models. Retrieved from https://www.theanalysisfactor.com/six-differences-between-repeated-measures-anova-and-linear-mixed-models

The Analysis Factor. (2008–2018b). The analysis factor. Retrieved from www.theanalysisfactor.com

The College of Physicians of Philadelphia. (2018, Januar 25). Do vaccines cause autism? Retrieved from https://www.historyofvaccines.org/content/articles/do-vaccines-cause-autism

The Lancet (Editors). (2010). Retraction: Ileal-lymphoid-nodular hyperplasia, non-specific colitis, and pervasive developmental disorder in children. *The Lancet, 375,* P445.

Tuskegee University National Center for Bioethics in Research and Health Care. (2018). About the USPHS Syphilis Studt. Retrieved from http://tuskegeebioethics.org/about-the-usphs-syphilis-study

US Department of Health and Human Services. (2016, March 3). International Compilation of Human Research Standards. Retrieved from https://www.hhs.gov/ohrp/international/compilation-human-research-standards/index.html

US Department of Health and Human Services. (2017). Presidential Commission for the Study of Bioethical Issues. Retrieved from https://bioethicsarchive.georgetown.edu/pcsbi/node/851.html

United States Department of Agriculture. (1966, August 24). The Animal Welfare Act: Public Law 89-544 Act of August 24, 1966. National Agricultural Library.

Retrieved from https://www.nal.usda.gov/awic/animal-welfare-act-public-law-89-544-act-august-24-1966

van Belle, G. (2008). *Statistical rules of thumb.* Hoboken, NJ: Wiley.

Vanderbilty University. (2018). REDCap. Retrieved from redcap.vanderbilt.edu

Vidoni, E., Johnson, D. K., Morris, J. K., Van Sciver, A., Greer, C. S., Billinger, S. A., ... Burns, J. M. (2015). Dose-response of aerobic exercise on cognition: A community-based pilot randomized controlled trial. *Public Library of Science, 10*(7), e0131647.

Wechsler, D. (2008a). *Wechsler adult intelligence scale-fourth edition: Technical and interpretive manual.* San Antonio, TX: NCS Pearson.

Wechsler, D. (2008b). *Wechsler adult intelligence scale-fourth edition: Administration and scoring manual.* San Antonio, TX: NCS Pearson.

Weindling, P., von Villiez, A., Loewenau, A., & Farron, N. (2015). The victims of unethical human experiments and coerced research under National Socialism. *Endeavour, 40,* 1–6.

World Medical Association (1964–2013). *WMA declaration of Helsinki: Ethical principles for medical research involving human subjects.* Helsinki, Tokyo, Venice, Hong Kong, Somerset West, Edinburgh, Washington DC, Seoul, Fortaleza: Author.

Yaffe, K., Falvey, C. M., & Hoang, T. (2014). Connections between sleep and cognition inolder adults. *Lancet Neurology, 13,* 1017–1028.

Young, R. K., & Veldman, D. J. (1981). *Introductory statistics for the behavioral sciences* (4th ed.). Orlando, FL: Holt, Rinehart and Winston.

Zhu, J., & Chen, H.-Y. (2011). Utility of inferential norming with smaller sample sizes. *Journal of Psychoeducational Assessment, 29,* 570–580.

Index